Renewal and the Current Air Defense Mission

David Rubenson

Robert Everson

Robert Weissler

Jorge Muñoz

Appendix by Donald Mitchell

Prepared for the
United States Army

Arroyo Center

RAND

For more information on the RAND Arroyo Center, contact the Director of Operations, (310) 393-0411, extension 6500, or visit the Arroyo Center's Web site at http://www.rand.org/organization/ard/

This report documents the findings of a project related to the renewal of the McGregor Range. The McGregor Range is one of six
military land parcels that in 1986 were "withdrawn" from the public
domain for 15 years by Public Law 99-606, the Military Lands Withdrawal Act. These parcels comprise nearly 30 percent of the Department of Defense's 25 million acres. They will revert to the public
domain in 2001 unless Congress passes new legislation. The McGregor Range comprises nearly 700,000 of Fort Bliss's 1.12 million acres.
The Fort Bliss garrison is adjacent to El Paso, Texas, but the McGregor Range is located entirely in New Mexico.

A decision to renew the withdrawal will depend on the military need
for the land and congressional interest in the nonmilitary uses that
still occur on McGregor. This document discusses the military uses
of the land and their compatibility with ongoing nonmilitary uses. It
should allow Congress, the public, and the Army to gain a clearer
understanding of the reasons for maintaining the land in the military
system. Alternatives to the type of withdrawal specified in PL 99-606
are also discussed.

More than half of all military land is classified as "withdrawn." To
date this concept has had few practical policy implications. However, the changing politics of land use has increased scrutiny on the
uses of all public lands. We can expect the concept of "withdrawn"
lands to garner increased political attention. While this document
focuses on the McGregor Range, it should be of interest to a wider
audience concerned with Army land management and those responsible for legislation affecting public lands. The discussion of the

relationship between the McGregor Range and the nearby White Sands Missile Range should make the document of interest to those concerned with joint use across different military installations.

The work was sponsored by Dr. Andrew Vliet, director of the McGregor Renewal Office at Fort Bliss. The research was conducted in the Force Development and Technology Program of RAND's Arroyo Center, a federally funded research and development center sponsored by the United States Army.

CONTENTS

Preface   iii

Summary   vii

Abbreviations   xiii

THE BRIEFING   1

Appendix: LEGISLATIVE HISTORY OF
    PUBLIC LAW 99-606   91

The U.S. Department of Defense (DoD) manages 25 million acres of federal land. Most of it was assigned when there were few competing uses and few questions about DoD's need for land. But population growth, suburban sprawl, environmental concerns, and new categories of recreational use have changed the situation. The DoD is increasingly being asked to justify its land holdings and determine what can be returned to the public domain.

The issue has immediate policy implications because 16 million DoD acres are classified as "withdrawn" public land. This is typical for military land in the West, and the term implies a congressional promise to return the land to the public domain when it is no longer needed for military purposes. The urgency of the promise depends on the specific legislation that has withdrawn the land. Some has been withdrawn in perpetuity, while six major ranges, comprising nearly 30 percent of DoD's land, were withdrawn in 1986 for only 15 years under the Military Lands Withdrawal Act (Public Law 99-606).

Unless Congress passes new legislation, the 99-606 lands will revert to the public domain in 2001. The purpose of this report, written as an annotated briefing, is to evaluate the military need for one of the six ranges: the Army's McGregor Range in southern New Mexico. McGregor's 700,000 acres comprises more than half of Fort Bliss's 1.12 million acres. Sections of McGregor are also used for cattle grazing and other nonmilitary purposes along with the military uses. The Bureau of Land Management (BLM) manages the cattle-grazing program subject to Army access rules.

INTRODUCTION

This report considers four policy questions:

- Is the McGregor Range a critical Army-wide priority?

- How intensely does the military use the McGregor Range?

- Are military and nonmilitary uses balanced effectively, and what could change that balance?

- Is it possible to transfer McGregor activities to the adjacent White Sands Missile Range (WSMR), which has more land and declining activity?

These questions were chosen because low military usage rates, competing nonmilitary uses, declining activity levels at WSMR, and the slowness in initiating the McGregor renewal process have been cited as reasons to return the range to the public domain. Each of these issues is complicated by U.S. Air Force plans to build a new bombing range on McGregor for German air force units based at nearby Holloman Air Force Base.

Our discussion defines three distinct parts of the McGregor Range: the Tularosa Basin south of New Mexico Highway 506, the Otero Mesa, and the land north of Highway 506. The division is a product of boundaries established by the highway and where nonmilitary uses occur. Nonmilitary use occurs only on Otero Mesa and the region north of Highway 506. We present an alternative division based on geological characteristics. This consists of the Tularosa Basin (north and south of Highway 506), the Otero Mesa, and the Sacramento Foothills.

Using the four policy questions and the three regions, we present a policy matrix that is the organizing paradigm for the briefing.

THE FOUR POLICY ISSUES

We discuss each policy question for each of the three sections of the range. Land use data from McGregor's Range Control Office reporting low usage on the Otero Mesa failed to include missile safety fans in the usage data. Safety fans are the buffer areas needed to ensure that debris or off-course missiles do not cause safety concerns. Very-low-altitude air operations also add to the unmeasured usage of Otero Mesa and the area north of 506. We conclude that usage in these two areas is not intense, but it is significantly larger than early Army data indicated.

Next we consider the Army-wide role of Fort Bliss and McGregor Range. Realignments in the 1990s have made Fort Bliss the Department of Defense's center for air defense. Nearly $20 million of new fiber optic lines have been laid throughout the range to support this mission. This is an investment that is tied to the land. We also discuss the Roving Sands training exercise, which is DoD's only truly joint (and international) training event. A large Roving Sands exercise takes place every other year, and units are stationed throughout the range. The density of units is already higher than what is ideally encountered in realistic tactical situations; the loss of the Otero Mesa would further degrade the utility of the exercise.

McGregor's role in supporting Fort Bliss's national air defense mission fulfills a critical part of national military strategy. The Army may have been slow to fund the renewal for a variety of reasons. One possibility is the mismatch between Fort Bliss's broad air defense mission and the major command structure of the Army. Commands fund installation obligations, and Fort Bliss is part of the Training and Doctrine (TRADOC) command. However, Fort Bliss houses units from a diverse mix of commands, leaving TRADOC with the bulk of the bill.

But could this mission be moved to WSMR, with its larger land area and declining levels of activity? Roving Sands would be affected adversely by east-west geographical limitations at WSMR, but we conclude that there is probably adequate available land and airspace to relocate other activities from Otero Mesa and the region north of Highway 506. However, WSMR's financial structure and operating tradition are completely inconsistent with the training that occurs on McGregor. Unless DoD moves to alter those policies, it is impossible to contemplate such a transfer.

We discuss the nonmilitary missions on Otero Mesa and the area north of Highway 506. With respect to the condition of the land, the only major ecological damage has occurred near cattle watering holes. We then describe four Bureau of Land Management (BLM) policy objectives for nonmilitary use. Current nonmilitary use is limited to the grazing of 2,500 cattle and the issuance of approximately 1,500 recreational permits each year (for the entire installation of 1.12 million acres). Given this low level of use, and a military use consisting largely of missile safety fans, we find no fundamental conflict to dual use on the Otero Mesa and the area north of Highway 506. Past conflicts have been related to managerial procedures rather than fundamental resource constraints.

SENSITIVITY ANALYSIS

Our first sensitivity analysis involves an alternative division of the range consisting of the Otero Mesa, the Tularosa Basin (both north and south of Highway 506), and the Sacramento Foothills. The foothills are limited to the northernmost portion of the range, and none of the area is intensively utilized.

We then consider U.S. Air Force plans to build a new bombing range on Otero Mesa for German air force pilots located at nearby Holloman Air Force Base. Although it constitutes less than 2 percent of the nonmilitary use area, the range violates BLM's goal of confining "hard" military use to the Tularosa Basin. Under 99-606, the military has the authority to implement projects that reduce nonmilitary use. We also present an appendix that reviews the legislative history of 99-606 and illustrates Congress's intent to allow the military this discretion. A new Otero Mesa range appears to be the best military option. However, overcoming some of the institutional obstacles in the way of using WSMR for Army training would make more feasible the option of using existing ranges. This option would be further enhanced by small changes in the German air force's requirements.

Although McGregor renewal could be justified by today's air defense mission alone, the Otero Mesa bombing range strengthens the military importance of the McGregor Range. For some it raises questions about the viability of nonmilitary uses. The 15-year withdrawal period is long relative to the time it takes to implement projects that could erode nonmilitary uses. We identify five legislative options with varying renewal periods and requirements for renewal.

CONCLUSIONS

We conclude by presenting answers to the four policy questions with which we began:

- Is the McGregor Range a critical Army-wide priority?

  We find that Fort Bliss has a critical role as the national center for air defense and that McGregor Range is essential for fulfilling that role.

- How intensely does the military use the McGregor Range?

  There is intense military use in the Tularosa Basin. There is a moderate level of low-impact use on the Otero Mesa and the Sacramento Foothills.

- Are military and nonmilitary uses balanced effectively, and what could change that balance?

  There are no fundamental obstacles preventing dual use for today's mission. The military uses have low impact and the nonmilitary uses are small. Ongoing Army efforts to respond to outside users will eliminate most conflicts. The new bombing range on Otero Mesa will not affect this, but it has created political concerns about the future. However, Congress clearly intended to give the military services the discretion to initiate projects that would affect dual use. Nevertheless, the dual-use priority has discouraged Army interest in utilizing the Otero Mesa for new military applications.

- Is it possible to transfer McGregor activities to the adjacent White Sands Missile Range (WSMR), which has more land and declining activity?

  There is probably sufficient land and airspace to transfer most activities on Otero Mesa and the area north of Highway 506 to WSMR. The situation may not be ideal for the Roving Sands exercise and would force other Fort Bliss units to travel greater distances. Substantial DoD policies and procedures currently prevent consideration of this transfer.

We also recommend that DoD establish a regional planning commit-tee to facilitate joint use of the Holloman/WSMR/Bliss complex.

ABBREVIATIONS

ACC	Air Combat Command
ACR	Armored Cavalry Regiment
ADA	Air Defense Artillery
AFB	Air Force Base
AMC	Army Materiel Command
AMRAAM	Advanced Medium-Range Air-to-Air Missile
ATACMS	Army Tactical Missile System
AUM	Animal Unit Monthly
BLM	Bureau of Land Management
CONUS	Continental United States
DCD	Directorate of Combat Development (Air Defense School)
DoD	Department of Defense
EIS	Environmental Impact Statement
FLPMA	Federal Land Policy and Management Act
FMSD	Field Mission Simulation Digital
GAF	German Air Force
GIS	Geographic Information Systems
LEIS	Legislative Environmental Impact Statement

MLWA	Military Lands Withdrawal Act
MRTFB	Major Range Test Facility Base
MTR	Military Training Route
NEPA	National Environmental Policy Act
NWRAA	National Wildlife Refuge Administration Act
OPTEMPO	Operating Tempo
OSD	Office of the Secretary of Defense
PL	Public Law
RPV	Remotely Piloted Vehicle
SHORAD	Short-Range Air Defense
TEXCOM	Test and Experimentation Command
TFR	Terrain Following Radar
THAAD	Theater High Altitude Air Defense
TRADOC	Training and Doctrine Command
UTTR	Utah Test and Training Range
WSMR	White Sands Missile Range

Chapter One

THE BRIEFING

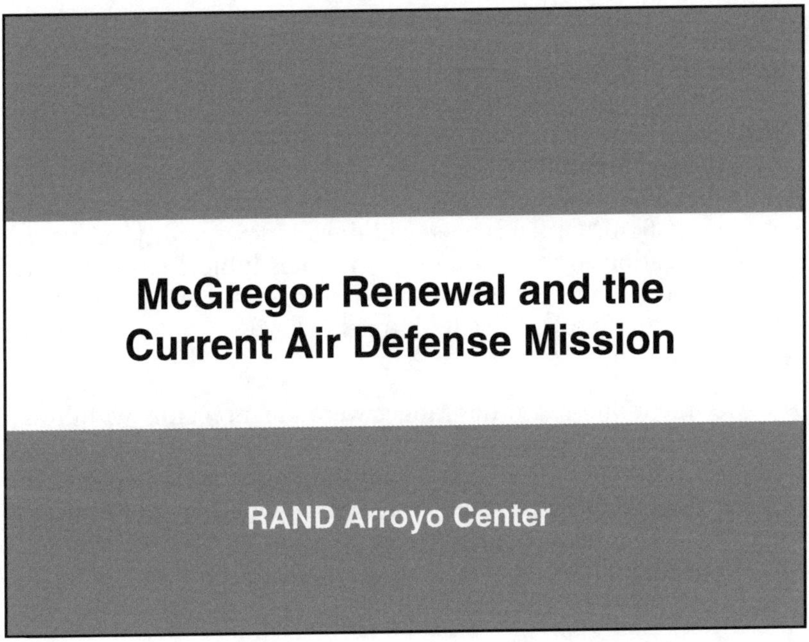
McGregor Renewal and the
Current Air Defense Mission
RAND Arroyo Center

Figure 1

This report, written as an annotated briefing, describes the military missions conducted on the Army's McGregor Range and places these activities in the context of an upcoming legislative discussion on the future of this land. The McGregor Range comprises almost 700,000 acres within Fort Bliss's 1.12 million acres. The main cantonment area of Fort Bliss is adjacent to El Paso, Texas, but the McGregor Range is entirely in New Mexico. Fort Bliss's Dona Ana range is also in New Mexico.

Existing legislation implies that Congress will need to determine by 2001 if the McGregor Range should remain under military management. This decision involves balancing national security needs against the nonmilitary use values associated with the intrinsic natural and cultural resources. Obviously there is no direct means of comparing the "worth" of these competing values, and the decision is appropriately left with political leadership. The purpose of this report is to enhance congressional, military, and public understanding of the military and nonmilitary issues at stake.

Most of the McGregor Range is classified as "withdrawn" public land.[1] This is typical for military land in the West, and the term implies a promise to return the land to the public domain when it is no longer needed for military purposes. The urgency of the promise depends on the specific legislation that has removed the land from public use.[2] Some land has been withdrawn in perpetuity, relegating the classification to little more than a philosophical reminder that the public retains title. The Dona Ana range on Fort Bliss and the nearby White Sands Missile Range (WSMR) are examples of withdrawals in perpetuity.

The McGregor and five other ranges were most recently withdrawn in 1986 for only 15 years under the Military Lands Withdrawal Act (Public Law 99-606). Congress established military activities as the primary use and gave the military authority to exclude public access for safety or national security concerns. But 99-606 also contains provisions for public use, and Congress made the Secretary of the In-

---

[1]The range consists of 608,385 acres of "withdrawn" public land, 71,083 acres of Army fee-owned land, 1,360 acres of Army fee-owned land within the Lincoln National Forest, 920 acres of previously state-owned land, and 18,004 acres of cooperatively used land within the Lincoln National Forest. The fee-owned land is fragmented and dispersed within the withdrawn land and has traditionally been assumed to be of limited military value by itself. However, new concepts such as the Army After Next, which may involve rapid insertion of light forces in a nontraditional battle front, could make small fragmented parcels a valuable training resource.

[2]Public Law 85-337, the Engle Act, requires that Congress approve any withdrawal greater than 5,000 acres for defense purposes. The Bureau of Land Management can authorize the withdrawal of smaller parcels but must review their status periodically.

terior responsible for managing the land in a manner consistent with the Federal Land Policy and Management Act (43 U.S.C. 1701), or FLPMA; the exception is priority for military activities even when those activities are inconsistent with FLPMA.[3] Congress dealt with the inconsistency of military priority and Interior Department management by directing the two department secretaries to conclude a Memorandum of Understanding (MOU). The McGregor MOU was completed in March 1990, and it gives the Interior Department's Bureau of Land Management (BLM) responsibilities for managing some nonmilitary uses, when such use is consistent with military missions and access control. The primary nonmilitary use is cattle grazing. This program is managed by the BLM.

Public Law 99-606 also directs the military services to prepare a draft Environmental Impact Statement (EIS) if they seek to renew withdrawals after 2001. The law requires that a draft legislative EIS (LEIS) be prepared by November 6, 1998.[4] Since legislation is needed for a renewal, the LEIS requirement is better seen as a congressional request for "good faith" by the military, rather than as a legal requirement. Congress could pass renewal legislation even if the military fails to issue the LEIS.

The short 15-year term and the specificity of the 1986 legislation are indicative of the changing political conditions for military land use. Previously, the low population density in the West dampened debate about military access to land. Some of the 99-606 lands had previously been withdrawn for 25 years, and there were even periods when the Army managed the McGregor Range without a legal basis.[5] However, the more restrictive 1986 legislation was the product of

---

[3]PL 99-606 is confusing on the relative authority of the Secretaries of Defense and Interior. Our interpretation of the legislative history is presented in the appendix and indicates that Congress hoped the two agencies could work together but gave DoD final authority to exclude nonmilitary uses.

[4]Fort Bliss published a draft LEIS in November 1998.

[5]The McGregor Range was first used for military purposes in the late 1940s but not formally withdrawn for a 20-year term until 1957. The Army managed the land without a formal legal mandate from 1977 to 1986. We should also note that the renewal period is not a single variable that perfectly correlates with growing competition over public land use. The California Desert Protection Act recently made the withdrawal of China Lake perpetual as part of a politically complex tradeoff in legislation that reorganized a significant part of federal holdings in the Mojave Desert.

emerging trends. Urban growth has created new categories of recreational land users. Hikers, hunters, off-road vehicle groups, preservationists, and others compete with the traditional ranching, forestry, and mining interests that long dominated public land use. These constituencies are well represented in local political discussions and increasingly by lobbying organizations in state capitals and Washington D.C. While much of public land still appears to be empty, there is now a fierce political competition for the use of these resources.

While congressional support for national defense remains strong, any military request for land is now carefully scrutinized for both its military requirements and its impact on alternative uses. This has been dramatically illustrated by the so-far unsuccessful efforts to expand the Army's National Training Center at Fort Irwin and the 15-year process to acquire a mere 12,000 acres near the Air Force's Mountain Home Air Force Base. Although these installations are located in areas with extremely low population density, requests for expansion met stiff political opposition and lingered for more than a decade.[6]

The renewal of already withdrawn lands does not affect the existing boundaries of public use lands and as a result has generally had less dramatic political impact than efforts to expand military land. However, any opportunity to recategorize federal lands is becoming a significant political event. McGregor is one of the 99-606 parcels that could create some controversy.[7] A U.S. Air Force plan to develop a new bombing range on McGregor for German air force (GAF) pilots based at nearby Holloman Air Force Base has aroused significant local controversy. This has reawakened memory of local ranching interests who have long held that the lands in the area (particularly the nearby White Sands Missile Range) were unfairly taken and that

---

[6]Both of these initiatives have been in process for more than a decade. The original Air Force proposal for a 1.5-million-acre expansion was reduced several times and became a 12,000-acre proposal. Congress approved the necessary legislation in October 1998. At the National Training Center, the Army had abandoned its original proposal for additional maneuver area in favor of a land area that would provide space for logistical support. But recent discussions suggest that after a 15-year effort, the original proposal may yet be viable.

[7]Concerns about endangered species at Goldwater and other land initiatives at Fallon Naval Air Station have aroused public interest in these withdrawals.

appropriate compensation was never offered. Although many McGregor ranchers sold voluntarily, one rancher did resist. The memory of this incident has been firmly engrained in the region through Edward Abbey's novel *Fire on the Mountain* (New York: Avon Books, 1992), which presents the story of a rancher's battle to resist federal efforts to confiscate land. The local office of the Bureau of Land Management has suggested that the scope of the renewal be reduced.[8]

As a result of these events, several questions have emerged related to the need for the military to keep the land. One question results from a perceived lack of military use on McGregor. A second concern is that the proposed bombing range may preclude existing nonmilitary uses. Some believe the military has more than enough land in the region and could easily transfer McGregor activities to nearby installations. Finally, the Army's slowness in initiating the renewal process may suggest that McGregor's mission does not represent a critical Army priority.

As such, this briefing is aimed at supporting a public, military, and congressional review of the renewed process by answering the following questions:

- Is the McGregor Range a critical Army-wide priority?

- How intensely does the military use the McGregor Range?

- Are military and nonmilitary uses balanced effectively, and will that balance change?

- Is it possible to transfer McGregor activities to the adjacent White Sands Missile Range (WSMR), which has more land and declining activity?

We note that Fort Bliss contains 10 percent of the Army's 12 million acres of land and is the largest Army facility used for military training (as opposed to weapons testing). With new military techniques requiring additional space and the need to conduct activities with

---

[8]A memorandum contained in a publication by the Las Cruces District Office of the Bureau of Land Management titled *McGregor Range Issues*, March 1998, states that the local BLM opposes renewal of a significant portion of the range and asserts that Secretary of the Interior maintains the same position.

increased environmental sensitivity, the Army finds itself facing a long-term challenge in meeting requirements for training land. The unique size of Fort Bliss might be an important factor in solving this dilemma. However, such a role might require a significant realignment of forces or a change in the relationship between units and home stations. This longer-term "strategic" role for McGregor Range will be the subject of another report that examines overall Army land management strategy. This report focuses on Fort Bliss's current mission and other currently planned activities.

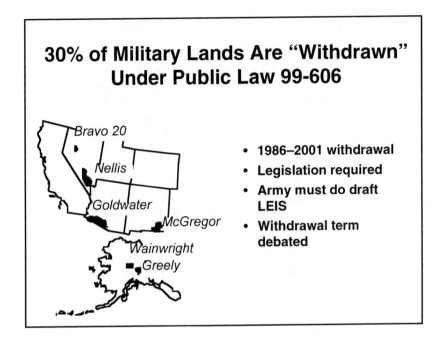

Figure 2

Figure 2 summarizes the policy issues posed by the expiration of the 99-606 withdrawals and shows the locations of the affected lands. The six parcels include Fort Greely, Fort Wainwright, the Nellis Range, the Goldwater Range (Arizona), the Bravo 20 Range at Fallon Naval Air Station (Nevada), and the McGregor Range. These lands comprise approximately 30 percent of DoD's 25 million acres of land. They are relatively isolated and in regions with low population density.

The renewal of these lands is an important legislative priority for the Department of Defense. The three military services have initiated draft Environmental Impact Statements (EIS) required by the 1986 legislation as a prerequisite for renewal. Because Congress must write new legislation to renew the withdrawals, failure to complete the draft legislative EIS (LEIS) would show a disregard for the 1986 legislation, but it would not preclude Congress from renewing the withdrawals. Since Congress is the ultimate decisionmaker, the LEIS does not provide the same mechanisms for injunctive relief as other EIS processes do.[9]

While each parcel evokes a different set of issues, the length of the withdrawal and the balance between military use and nonmilitary use have traditionally dominated discussion. The Alaska sites are isolated and have aroused few public comments. Scoping meetings for the Bravo 20 LEIS were sparsely attended, but other Fallon land management issues might ultimately be connected (politically) to the renewal. The Navy is attempting to withdraw 135,000 additional acres in a separate legislative initiative that has drawn criticism from both citizen groups and the state of Nevada. There has also been a long history of controversy associated with the proximity of the Bravo 16 bombing range to the town.

Goldwater and McGregor have to date probably raised the most discussion, though neither renewal appears in jeopardy. Goldwater contains a wildlife refuge within its borders, making access, impact, and management a source of continuing debate.[10] As noted above, ongoing nonmilitary uses complicate the McGregor renewal. Controversy seemed to be temporarily reawakened by recent U.S. Air Force proposals to build a new bombing range on McGregor for use by German air force pilots based at Holloman. Ranchers complain about the potential noise impacts, and others see the proposal as the first step toward elimination of nonmilitary uses.[11]

---

[9]Even if 99-606 did not specify an EIS as a prerequisite for renewal, language in the law specifying that the National Environmental Policy Act (NEPA) applies to actions taken on the withdrawn land *might* imply the need for EIS in any case. But since Congress is the ultimate decisionmaker, it could address any flaw in a traditional NEPA document in the renewal legislation.

[10]See the World Wide Web site http://www.rama-usa.org/goldwat.htm for Wilderness Society comments on the Goldwater renewal.

[11]Complaints about noise are associated less with the range than with the Military Training Routes (MTRs) that are used to access the range.

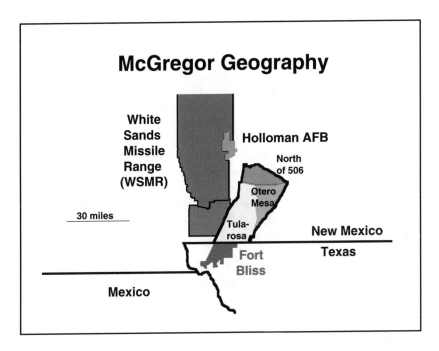

Figure 3a

Figure 3a illustrates the location of the McGregor Range relative to the remainder of Fort Bliss, the White Sands Missile Range, Holloman Air Force Base, and the major state boundaries in the area. WSMR is part of the U.S. Army Materiel Command (AMC), and Holloman is part of the U.S. Air Force's Air Combat Command (ACC). Fort Bliss is part of the Army's Training and Doctrine Command (TRADOC). The McGregor Range is outlined with a heavy line and constitutes 700,000 acres of Bliss's 1,120,000 acres. It lies entirely in New Mexico. The main cantonment area of Fort Bliss is shown in dark green and lies entirely in Texas. Fort Bliss's Dona Ana Range (purple) nearly connects the main cantonment area of Fort Bliss (through the McGregor Range) with WSMR. State Highway 54 is the boundary between the McGregor and Dona Ana ranges.

It is interesting to note that with the Fort Bliss main cantonment area adjacent to El Paso, most of the economic spillover occurs in Texas. The Army was initially slow to provide funding for the LEIS process, and it was only the actions of Texas Senator Phil Gramm that led to the process being initiated.[12]

Figure 3a divides the McGregor Range into three regions: the portion of the Tularosa Basin south of New Mexico Highway 506, the entire region north of Highway 506 (comprising part of the Tularosa Basin and the Sacramento Foothills), and the Otero Mesa. The Tularosa Basin south of Highway 506 is a long flat plain characterized as Chihuahuan desert grassland and creosote bush scrubland. The region north of 506 comprises an extension of the Tularosa basin in the northwest and the Sacramento Foothills in the northeast corner of the region. The Otero Mesa is also a flat plain characterized as Chihuahuan desert grassland that is elevated from the Tularosa by an escarpment of about 1,000 feet.

---

[12]The FY 97 Senate version of the budget was supplemented by amendment 4582 introduced by Senator Gramm. The amendment stated, "Of the funds appropriated in Title II of this Act, not less than $7.1 million is available to perform the environmental impact and associated baseline studies necessary to prepare an application for renewal of use of the McGregor Range at Fort Bliss, Texas." The McGregor Range lies entirely in New Mexico.

View Across Otero Mesa

Chihuahuan desert grassland typical of the
mesa near proposed bombing range target site

Figure 3b

Figure 3b shows a typical view of the Otero Mesa. The division in Figure 3a is the product of the nonmilitary uses associated with each area. The Tularosa Basin, south of 506, has no nonmilitary uses. The areas north of 506 and the Otero Mesa were withdrawn along with the rest of McGregor in 1957 and are used for active grazing as well as some hunting, recreational use, minor commodity sales, and aesthetic value. The area north of 506 contains the approximately 10,000-acre Culp Canyon Wilderness Study Area. It is used for low-impact recreational activities such as hunting, hiking, and camping. The Otero Mesa, including the escarpment, contains about 4,000 acres of native black gamma grass habitat and is used as study areas for ecology researchers.

An alternative division of the range could be made by dividing the base into the Otero Mesa, the Sacramento Foothills, and the entire Tularosa Basin (both north and south of Highway 506). This division better corresponds to natural geologic boundaries. It does not, however, distinguish the portion of the Tularosa Basin north of Highway

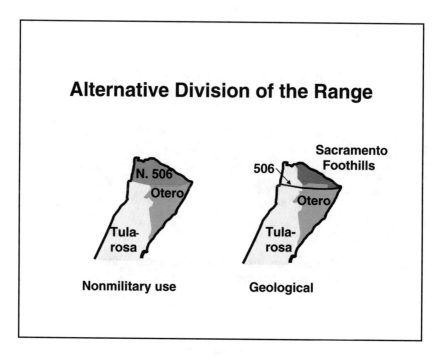

Figure 3c

506 that is used for cattle grazing. Figure 3c contrasts this division with the one shown in Figure 3a.

The grazing, recreation, hunting, and other nonmilitary uses of the McGregor Range are managed by BLM, as arranged in a Memorandum of Understanding between the Army and the BLM. The Army grants access permits for both BLM personnel and nonmilitary users.

Arguments for Limiting Renewal?

- **Light use?**
- **Nonmilitary uses?**
- **Access to White Sands?**
- **Lack of critical Army-wide role?**

USAF planning bombing range for German air force

Figure 4

Figure 4 summarizes the four issues that correspond to the questions that shape this study. These are the arguments that have been made for limiting or eliminating the renewal. They have been made by members of the public at LEIS scoping, by the BLM, and by some members of the Army itself.[13] They are similar to arguments presented at the 1986 congressional hearings on PL 99-606.[14] The primary issue is the desire of nonmilitary users to access and use the McGregor Range. Since these activities already occur on McGregor, the policy issue is the extent of nonmilitary use, the desirability of the

---

[13]Made at the public scoping meeting and in writing. See *McGregor Range Issues,* January 1997, published by the Las Cruces Office of the BLM. See also *Scoping Summary, McGregor Range Land Withdrawal Renewal,* February 1998, for actual transcripts of comments from government agencies and the public. As will be discussed later, preliminary Fort Bliss range utilization data were circulated within the Army and led some to suggest that low utilization might indicate a lack of need for the range.

[14]See U.S. Army, *Preliminary Draft Land Use Study, McGregor Range, New Mexico Land Withdrawal Renewal, August 1997,* for a good summary of the 1986 hearings.

existing dual-management structure, and whether changing military requirements might degrade them.

The desire to return the land to the public domain is enhanced by any perception that parts of the range are only lightly utilized and hence of minimal military value. As will be discussed in succeeding figures, this perception is due in part to the Army's own approach to summarizing range usage data.

The argument for limiting the renewal is bolstered by the presence of the White Sands Missile Range adjacent to Fort Bliss. Many of WSMR's launch facilities are in the extreme southern area of the test range, so movement from Bliss's cantonment area is not pro-hibitively far. WSMR is the Army's largest installation. It is widely perceived as having declining levels of activity, and its missile testing mission has similarities to Fort Bliss's air defense mission. Both in-stallations have extensive launching facilities, restricted airspace, and missions with little heavy ground impact. This suggests that it would be easy to transfer McGregor activities to WSMR.

It is also true that the Army's Training and Doctrine Command was initially hesitant to pay for the environmental studies specified in 99-606 as a prerequisite for renewal. This might be interpreted as a lack of Army-wide interest in the mission occurring on McGregor. The LEIS process at McGregor began well after the processes were initiated at the Air Force's Goldwater and Nellis Ranges. After the money failed to be allotted through the normal command channels, only the actions of Senator Phil Gramm and a specific budget amendment allowed the studies to proceed.

Finally, we reiterate the plans to construct a U.S. Air Force bombing range on the Otero Mesa primarily for use by German air force pilots based at nearby Holloman Air Force Base. The proposal has caused significant controversy in the region, and the recent decision to move forward[15] will alter the potency of the arguments presented above: If the range is built on Otero Mesa as planned, the issue of limited military use is less credible. It also increases concerns about the long-term viability of nonmilitary uses.

---

[15]The U.S. Air Force issued a Record of Decision (ROD) on May 29, 1998. Construction has not yet begun.

Policy Matrix

Issue	Tularosa (South 506)	North of 506	Otero Mesa
Military use			
Nonmilitary use			
Easy exit (use WSMR)			
Army-wide role			

Implications for USAF/GAF bombing range?

Figure 5

Figure 5 illustrates the analytical framework that we will use to discuss the issues identified in Figure 4. As indicated, we will discuss these issues for each of the three geographical regions highlighted in Figure 3a—a division of the range based on nonmilitary use. We will consider the implications of the division represented in Figure 3b as a sensitivity analysis. We will also consider the bombing range on the Otero Mesa as part of the sensitivity analysis.

Arguments for Limiting Renewal?

Issue	Tularosa (South 506)	North of 506	Otero Mesa
Military use	High use	Low use	
Nonmilitary use	None	Grazing, conservation, and recreation	
Easy exit (use WSMR)	2 million adjacent acres with declining activity		
Army-wide role	Army hesitancy to fund renewal		

Counter Arguments?

Figure 6

Figure 6 places the arguments presented with Figure 4 into the policy matrix and poses the central question for the briefing: What are the counter arguments? Green depicts situations where the military would seem to have a strong basis for seeking renewal, and red shows the opposite situation. As mentioned above, a perception of light military use on Otero Mesa and north of Highway 506, along with nonmilitary uses, has led some to request that these regions be returned to the public domain. This argument is enhanced by the assumed ability to relocate limited activities to WSMR. The Army's own hesitancy in funding the McGregor renewal process has led to questions about the Army-wide value of the range, independent of usage levels.

The remainder of the briefing will be organized by the policy matrix. The four central policy questions correspond to the rows of the matrix. We will first address military usage rates (as indicated by the darkened horizontal rectangle) and then proceed to the Army-wide role, the potential to relocate to WSMR, and finally the nonmilitary uses. For each row we will provide additional background on the arguments for limiting renewal. We will then describe the arguments that must be made to convert the red box to green, or at least to orange, which indicates our judgment that there is a partially effective counter argument.

We conclude by examining two sensitivity cases. First we consider the impact of the division presented in Figure 3b. We then consider the impact of the bombing range on Otero Mesa. Although the Air Force has issued a formal Record of Decision, budgetary or legal issues could reopen the debate. We therefore include a brief discussion of the rationale and need for the range.

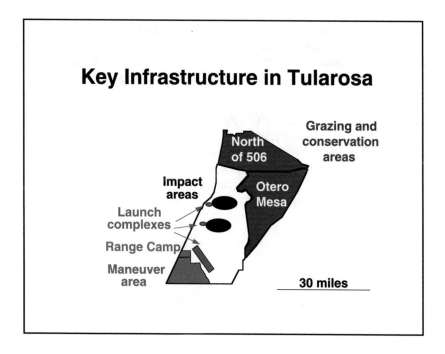

Figure 7

Figure 7 highlights one aspect of the first argument for limiting the renewal: that most military use and military infrastructure lies in the Tularosa Basin. The Otero Mesa and the area north of 506 have minimal military infrastructure and contain the BLM-managed grazing program and other nonmilitary uses. Figure 7 illustrates that the major missile launch complexes are located in the Tularosa Basin, south of Highway 506. The short-range complexes are in the center of the basin, and the long-range complex is in the south. The Range Camp, which contains the command and control for range activities, is on the southwestern edge of the basin. The major impact areas are located in the central part of the basin, meaning extensive cleanup efforts might be needed before this land could be returned to the public domain.

There is a Class C air-to-ground target in the northwest corner of the range within the area categorized as "north of Highway 506." This target is west of the Sacramento Foothills region highlighted in Figure 3c.

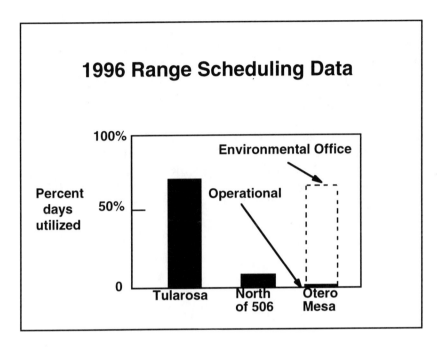

Figure 8

Figure 8 illustrates data derived from the Fort Bliss Range Control schedules.[16] The chart shows the percentage of days that military activity occurred in 1996 in the three geographical regions. The data show that the Tularosa Basin south of 506 is used frequently but that the other two regions are rarely used. The Army's own range data indicate that the only significant military user of the Otero Mesa was Fort Bliss's environmental office, which conducts extensive data gathering and surveys in order to comply with environmental regulations.

---

[16]The data shown in Figure 8 are taken from 1996 Fort Bliss Range control records. In actuality the three regions are divided into a total of 26 maneuver areas, and utilization for each is reported. The Figure 8 data represent a rough averaging for each of the three ranges and are only intended to represent a qualitative finding. 1996 is used as the base year in this analysis, as it is in the draft Environmental Impact Statement. Although, as will be discussed later, there has been some realignment of forces, long-term use patterns are not yet in place for the new forces. Use patterns for 1997 are not qualitatively different from 1996.

These data support the idea that the Otero Mesa and the region north of 506 could be returned to the public domain with minimal impact on the Fort Bliss military mission. The data were summarized in internal Army documents and led some within the Army to conclude that the Otero Mesa and the region north of 506 were not critical Army priorities.

We should note that the higher utilization of the area north of Highway 506 (as opposed to the Otero Mesa) is due largely to the western portion of this region, which contains the Class C target area. This will be discussed in greater detail in the sensitivity analysis where we consider alternative divisions of the range.

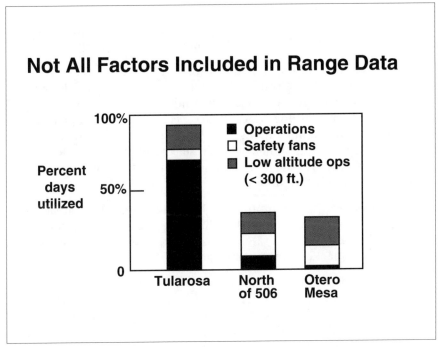

Figure 9

Figure 9 adapts the utilization data shown in Figure 8 into a more complete picture of military usage.[17] The key difference is the addition of missile safety fans and low-altitude aircraft overflights.

The Army's range control data only reported scheduled uses of the 26 maneuver areas or missile firing points. Data were reported as if Fort Bliss had a classic Army mission emphasizing ground operations. As will be discussed in the next section, Fort Bliss's central mission is air defense, and most air defense missile flights require a large secure area to ensure safety of personnel on the ground. It is always possible for a missile to misfire and land far from the intended target. De-

---

[17]The data again represent averaged values over the 26 maneuver areas. The Army's forthcoming *McGregor Range, New Mexico Land Withdrawal Renewal Legislative Environmental Impact Statement* describes the usage by placing each area into either a high (75–100 percent), moderate (50–75 percent), low (25–50 percent), or very low (0–25 percent) category. See Section 2 of the forthcoming document.

bris may also scatter over a large area. A missile firing requires large patches of land, even though most of that land is rarely affected.

Figure 9 also shows the utilization for low-altitude aircraft flying routes under 300-foot altitude. Technically it is not essential to own the land under utilized airspace. The military overflies vast areas of the public domain it does not manage. However, the frequency of the low-altitude flights implies that if this land were turned back to the public domain, the military would seek to maintain the floor-to-ground airspace over the land, and new nonmilitary users would be forced to contend with this impact.

Taken together, the low-altitude flights, safety fans, and the actual scheduled usage give a somewhat different picture of overall range usage. The range is moderately utilized but in a way that has minimal "hard" impact on events on the ground.

It is interesting to note that the Wilderness Society, Friends of Cabeza Prieta, and the Defenders of Wildlife have argued that military land used only for overflights and safety fans should be returned to the public domain.[18] Although there is no suggestion for how safety might be achieved, the proposal may highlight the necessity of developing and testing new approaches. Weapon system range is increasing, as are the costs of actual firings. In the future we will see fewer missiles launched, but each may need to overfly larger distances. Procedures for temporary military usage of larger areas, as an alternative to withdrawal or acquisition of smaller areas, may become increasingly desirable.

---

[18]Written comments submitted in response to the Barry M. Goldwater Air Force Range Renewal Legislative EIS, http://www.rama-usa.org/goldwat.htm.

Arguments for Limiting Renewal?

Issue	Tularosa (South 506)	North of 506	Otero Mesa
Military use	High use	Moderate use with light impact	
Nonmilitary use	None	Grazing, conservation, and recreation	
Easy exit (use WSMR)	2 million adjacent acres with declining activity		
Army-wide role	Army hesitancy to fund renewal		

Counter Arguments?

Figure 10

Figure 10 recasts the Figure 9 data in terms of the policy matrix. We have colored the military usage boxes for north of 506 and Otero Mesa orange, as these ranges are used more than Figure 8 indicated; still, it is difficult to characterize this land as intensely used. The data in Figure 9 show the percentage of days utilized for military purposes, even if the range was used only a small portion of a day. Many Fort Bliss personnel still felt there was time available to be scheduled. This was one of the primary reasons the Air Force chose the Otero Mesa for a new bombing range.

As indicated by the dark outline on the bottom row, we will now address McGregor's overall Army role, independent of the levels of usage. The need for examination is motivated by the Army's initial hesitancy to fund the EIS needed to file a renewal application. This suggests to some that the Army did not see renewal as a top priority.

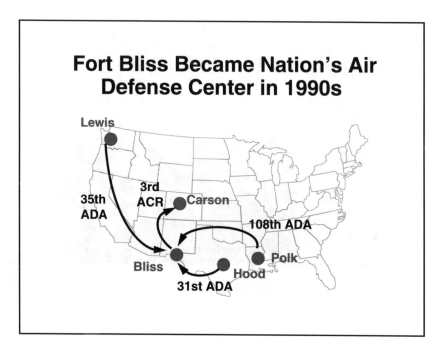

Figure 11

Fort Bliss is part of the Army's Training and Doctrine Command (TRADOC). This is the group of installations that house the Army schools where new soldiers are given basic training. The schools also develop advanced concepts and tactics for the individual military branches. Fort Bliss houses the Air Defense School. TRADOC does not include active war-ready units, though such units are at times stationed at TRADOC bases.

Fort Bliss's installation functions (including those of the McGregor Range) are funded through TRADOC accounts. Figure 11 indicates that units from Forces Command, the command that contains most of the Army's active units, were realigned to Fort Bliss in the mid-1990s. Corps-level and above air defense artillery (ADA) brigades were transferred to Bliss even though corps headquarters were located at the Forces Command posts of Forts Lewis, Hood, and Bragg. In addition, another Forces Command unit, the 3rd Armored Cavalry Regiment (ACR), was transferred out of Fort Bliss to Fort Carson.

The active units transferred to Fort Bliss constitute all domestically based active air defense units at corps level or above. Along with the 6th ADA Brigade, which is used primarily to support the Air Defense School function, all of the Army's CONUS-based long-range air defense is now located at Fort Bliss.

This statement is significant because Fort Bliss's role in the Forces Command structure has become as important as its role in the TRADOC structure. Nevertheless, funding for installation support activities comes largely through TRADOC channels.

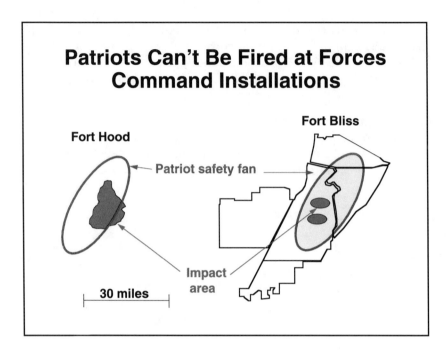

Figure 12

Figure 12 illustrates one of the primary reasons Fort Bliss was chosen as the nation's home for active air defense at the level of corps or above; it is the only installation in the training portion of the Army that is large enough to accommodate the firing of long-range or even medium-range air defense missiles. Figure 12 displays an overlay of the safety fan of a Patriot missile on Fort Bliss and on one of the largest Forces Command home stations, Fort Hood. Fort Hood was the home of the 31st ADA prior to the realignment illustrated in Figure 11. The Patriot, or missiles of significantly shorter range, could not fit within its confines. Figure 12 shows that it is necessary to have an installation that is both large and shaped in the elongated form of the safety fan. This accounts for the rectangular shape of WSMR, which was founded for developmental missile testing.

There is no Forces Command home station that could accommodate a Patriot safety fan. Fort Stewart, the largest home station, is 20 percent larger than Hood but also falls far short of the necessary dimensions. The 642,000-acre Fort Irwin is only 10 percent smaller than

McGregor but is not properly shaped to accommodate the fan. Whereas the longest dimension of Bliss/McGregor is approximately 70 miles, Fort Irwin is shaped more like a square.

Today's air defense missiles are increasingly expensive to buy and are fired less and less during training. A Patriot missile costs about $1 million. Actual firings are limited to the minimum number that will give units the confidence that skills learned during simulation can be replicated in battlefield situations. Given the relative ease of simulating air defense battles, as compared to ground scenarios, air defense training is already more dependent on a mixture of simulation and actual field exercises than other combat branches of the Army. This factor also acted to increase the rationale for transferring the active Patriot units to Fort Bliss. The TRADOC schools help develop and utilize simulations. Air defense simulations are practiced and developed at Fort Bliss.

TRADOC Only One Part of Bliss Mission

- **Classical TRADOC**
  - Air defense school
- **Intercommand**
  - Forces Command: All CONUS air defense brigades
  - TEXCOM (operational testing)
  - Extended range for WSMR developmental testing
  - **Interservice**
  - Marine school
- **International**
  - German school
  - Visiting foreign units

Figure 13

Figure 13 integrates the discussions in Figures 11 and 12 into an overall picture of the military mission at Fort Bliss. The classic TRADOC schoolhouse function comprises only a small portion of the installation's diverse air defense mission. However, TRADOC is responsible for paying most installation infrastructure and maintenance costs.

As discussed above, Fort Bliss is now the home for all Forces Command air defense units at corps level or above. It is also the home for the Test and Experimentation Command's (TEXCOM's) air defense operations. TEXCOM is an independent Army agency responsible for operational testing of new weapon systems. Operational testing occurs after a system has passed developmental test but has not yet been proved to be "user-friendly" for the training units. Independent operational testing is required by Congress. Fort Bliss is occasionally used as an extended range for developmental missile testing conducted at WSMR. The most recent and notable examples of this are the ATACMS (Army Tactical Missile Systems), which are launched at Bliss and land on WSMR.

Figure 13 also shows that Bliss has a role beyond the Army and is emerging as a national and international air defense center. The Marine Corps houses its air defense school at Bliss and uses the short-range air defense missile range on the Tularosa Basin. The German air defense school and German Patriot training activities are also housed at Bliss. There have been an increasing number of allied military units coming to Bliss to take advantage of the school and to conduct missile firing exercises.[19]

There is a significant mismatch between the diverse air defense missions performed at Bliss and the organization of the Army. This may partially account for the slowness in Army proponency for the renewal. The domestically based active Army is organized by major command with most activities occurring in either the Training and Doctrine Command, Forces Command, and Army Materiel Command. The Army is generally not organized by branch. For example, each major command is responsible for functions related to all branches. Forces Command contains the active armor units, TRADOC conducts basic armor training and develops armor doctrine, and AMC supports testing of new armor concepts. The same is true for infantry, engineers, and the other individual branches. However, for air defense, almost all of these functions are performed at Bliss. While the Army is organized by multibranch commands, the unique geographical features of Fort Bliss, and the unique way that field training and simulation are blended for air defense, have led the Army to organize Fort Bliss by branch rather than command function.

---

[19]On one visit to McGregor Range, we were told that Fort Bliss recently hosted Japanese and Dutch units for ADA training.

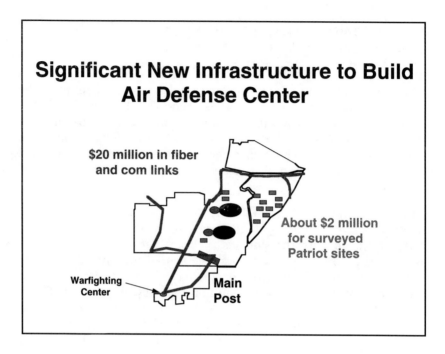

Figure 14

Although Figure 7 portrays the major infrastructure investment as being located on Tularosa, Figure 14 highlights important investments that have occurred on Otero Mesa since the major ADA units were transferred to Bliss. These investments are aimed at enhancing Fort Bliss's role as an international center for air defense. Most significant is the ongoing construction of a vast fiber optic network covering much of the McGregor Range, including Otero Mesa. During the last five years approximately $20 million has been spent on this system.[20] The first use of the fiber will be to ensure reliable communication with the security team responsible for keeping the public off the range during missile firings.

---

[20]The figure is based on a communication with the Directorate of Information Management (DOIM) at Fort Bliss. The figure represents a sum of several individual contracts.

More significantly, the fiber is expected to become a critical Fort Bliss training tool. As noted above, the cost of air defense missiles forces a significant amount of training to be accomplished through simulation. Units deployed in the field "fire" simulated Patriot missiles in response to electronically transmitted synthetic battlefields. Given that actual air defense battles are fought by reading electronic signals, this branch of simulation is already more realistic than more complicated efforts to simulate ground terrain and vehicles.

The synthetic environment is generated at the Fort Bliss Warfighting Center (main post) and is currently transmitted with either tactical links (airways) or through fiber optic lines. Conversion devices, Field Mission Simulation Digital (FMSD), convert these signals into synthetic radar signals that are fed directly into the Patriot system. Tactical transmission of the synthetic environment ties up significant manpower and communication equipment that would normally be used for other military functions. Fiber is simpler to use and more reliable, and it provides significantly more bandwidth. This allows for more diverse and complex simulation. Use of the fiber allows for larger exercises in which multiple ADA units can be engaged in larger overall battles. This allows for inclusion of different types of targets and a more complex and realistic environment.

The development of the fiber network link represents a major investment, tied to land and linked to the vision of Fort Bliss as a center of excellence for all air defense functions. The fiber network connectivity supports not only the ability to transmit signals but also the ability to have the reaction of the firing units transmitted back to the warfighting center for analysis and review. Fort Bliss currently has only six FMSDs, so the scope of simulations is still limited. But additional FMSDs are funded, and this will enable the development of more complex synthetic battles and increase dependence on the fiber system.

Figure 14 also highlights the approximately fifteen one-square-kilometer sites for Patriot deployments on Otero Mesa. These sites have been used since 1992 for the annual Roving Sands exercise. The units are confined to areas of one square kilometer[21] but are able to

---

[21]The impact of these activities to date is relatively limited to a very small portion of the one-square-kilometer areas near existing roads.

go through all the training required to deploy and simulate firing at simulated targets. An additional $2 million is currently being spent on environmental studies to place an additional 20–25 such sites on the mesa, primarily for use in routine ADA training.

Currently the fiber optic lines do not extend to these sites, but most of the distance between the sites and the warfighting center has already been covered with fiber. An eventual linkage would allow units to train in a synthetic battlefield while deployed in a realistic field setting.

It should also be noted that the fiber connection to WSMR is in place, though the switching to make the connection has not been activated. The linkage could enhance the effectiveness of test or training activities that must cross installation boundaries.

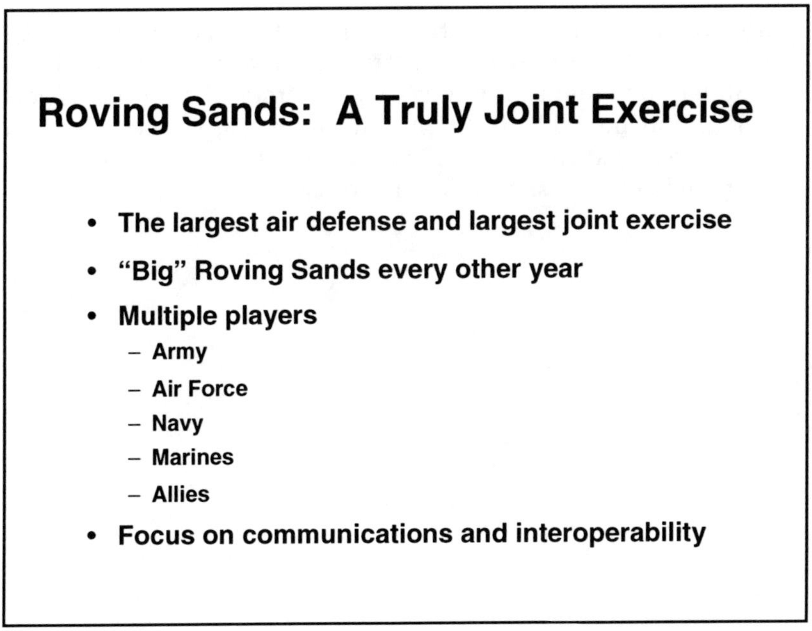

Roving Sands: A Truly Joint Exercise

- **The largest air defense and largest joint exercise**
- **"Big" Roving Sands every other year**
- **Multiple players**
  - **Army**
  - **Air Force**
  - **Navy**
  - **Marines**
  - **Allies**
- **Focus on communications and interoperability**

Figure 15

We have argued that physical size, the stationing of the Army's ADA units at corps level and above, the presence of international units, and the warfighting center make Fort Bliss a globally unique facility for air defense training. Figure 15 summarizes the Roving Sands training exercise, which takes advantage of these attributes. Roving Sands is arguably the most significant truly joint training exercise conducted by U.S. armed forces.

Roving Sands takes place every year in the late spring and probably represents the most important two weeks of training at Fort Bliss. The focus is on ensuring communications and interoperability among diverse air defense systems and organizations. The exercise also includes allied nations.

Roving Sands has grown in scope as air defense is increasingly seen as a more complex and important part of the nation's military posture. Every two years there is a large Roving Sands, with more units deployed across more of the Fort Bliss geography than in the alternating years. In 1997 the Army contributed parts of three ADA brigades, the Air Force over 50 aircraft, the Marine Corps its HAWK and Stinger battalions, and the Navy 13 aircraft based on carriers. German, British, and Dutch ADA units, as well as German F-4s and Tornados based at Holloman AFB, also participated. All the services contributed logistical support to the exercise.

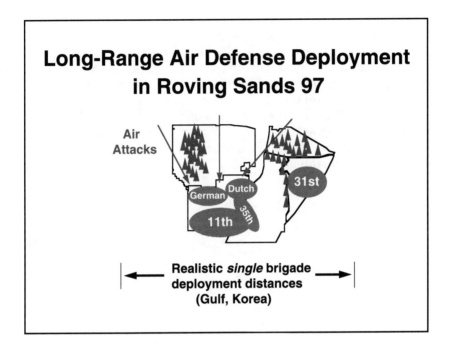

Figure 16

Figure 16 shows the approximate deployment regions (blue ovals) for the long-range ADA units in Roving Sands 97. The brown cones represent an approximate distribution of mountainous terrain. The 11th, 31st, and 35th Brigades were deployed along with Dutch and German ADA units. Air attacks originate from U.S. Air Force bases to the northeast of Fort Bliss, with the approach down the long corridor of floor-to-ceiling restricted airspace over WSMR. As can be seen in Figure 16, the 31st Brigade was deployed on Otero Mesa. The deployments were limited to presurveyed one-square-kilometer boxes. Deployment on Tularosa Basin was extremely limited in 1997 due to poor soil condition.

Those we interviewed at Forces Command and Fort Bliss spoke unanimously as to Roving Sands' value as an air defense training exercise. They cited it as the only opportunity to develop proper coordination with other units and the only opportunity to test air defense tactics at a theaterwide level.

Despite the value of the exercise, Figure 16 also shows that limitations on training space confine ADA deployments into significantly smaller areas than called for in operational scenarios. In Korea or the Persian Gulf, a single ADA brigade is deployed over a region as large as that being used for the five brigades deployed in Roving Sands 97. While dense ADA units present more effective defenses, individual units may be assigned far larger defensive zones in realistic scenarios. Thus even with the large expanses of Fort Bliss, Roving Sands is already space constrained.

Loss of Otero Mesa Would Damage Large Roving Sands

- **Trainers view exercise as providing excellent training, but . . .**

- **Unrealistic density**
    - Not an optimal test of interoperability
    - Air forces don't penetrate and are reluctant participants

- **Insufficient sites for realistic survivability techniques**

Figure 17

Figure 17 summarizes the military training benefits of the Roving Sands exercise and highlights the limitations posed by geographical constraints. Despite geographical limitations, there was consensus among Army trainers as to the training value of the exercise. The loss of Otero Mesa would exacerbate existing limitations posed by geography. To date this would only affect the large Roving Sands exercise every other year.

As indicated in Figure 16, distances are already significantly smaller than encountered in existing deployments. As a result, the radar coverages in Roving Sands have overly extensive and unrealistic overlaps and the exercise is not as realistic a test of interoperability as desired. More significantly, the current deployments produce an unrealistically dense air defense picture for attacking pilots, so the air forces typically do not try to penetrate. This reduces the training utility for pilots and has led some Air Force planners to question the value of the exercise. Finally, the lack of siting options for the Patriots inhibits the units from realistically implementing survivability training tactics requiring site rotation. Patriot units are limited to a small number of sites that become overly predictable.

The loss of Otero Mesa would exacerbate all of these problems. It would be possible to site the 31st ADA Brigade on the Tularosa Basin, but the result would be an even denser air defense formation. The soils in the basin are fragile and require rest periods such as occurred in 1997.

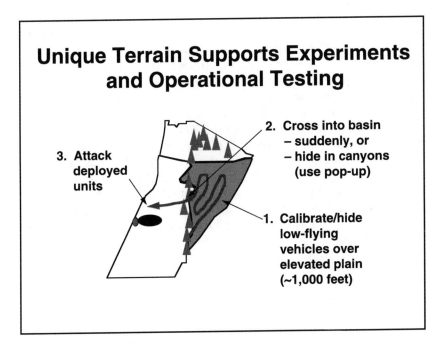

Figure 18

As a final note, Figure 18 highlights factors that make Otero Mesa critical to the fulfillment of the operational testing mission and to the warfighting experiments conducted within the Air Defense School's Directorate of Combat Development (DCD) and TEXCOM.

The unique terrain features of the Otero Mesa and the McGregor Range are exploited to develop tests that combine the need for precise measurement with the need for attacks that simulate the surprise and shock of real scenarios. This is accomplished by calibrating low-altitude threat vehicles as they fly over the flat mesa. Since the mesa is as much as 1,000 feet above the basin floor, the aircraft can be deliberately calibrated while remaining invisible to units deployed on the western edge of the Tularosa Basin. The threat air vehicles (aircraft, drones, helicopters, etc.) can then be brought westward across the basin. Test managers can allow the vehicles to appear suddenly over cliff-like structures, or they can bring them down canyon formations to hide the targets. Helicopters can implement "pop-up" tactics by flying down canyons that provide places to hide.

One of the most recent tests occurred in December 1997, in efforts to look at the AMRAAM missile in a ground-launched configuration. Several weeks of testing occurred to determine the effectiveness against drones, RPVs, slow fixed-wing aircraft, fast fixed-wing aircraft, and simulated cruise missiles. The target vehicle was launched from the western edge of the Tularosa Basin and directed eastward over the Otero Mesa. Calibration occurred over the mesa with tracking radars deployed there. The target flew at approximately 300 feet above the mesa floor so that it was invisible to both radar and the human eye from the SHORAD range at the western edge of the basin. The missile was then brought over the structure called the Pandejo wash, where it was visible but partially masked by terrain features. Intercept occurred as the missile flew low over the flat part of the Tularosa Basin.

Both the Directorates of Combat Development (part of the Air Defense School) and TEXCOM use the McGregor Range for testing. Several major tests or experiments are conducted each year. In 1996, which is our base year, TEXCOM and DCD conducted six major testing programs. These included LINEBACKER (a Stinger system mounted on a Bradley fighting vehicle), Patriot Advanced Configuration 2, the Integrated Target Acquisition System (ITAS), and the Joint Tactical Information Distribution System (JTIDS). FY 98 is likely to be a light year, but the third Patriot Configuration, the Forward Area Air Defense (FAAD) system, and the Tactical High Energy Laser (THEL) are scheduled for operational test in FY 99.

Arguments for Limiting Renewal?

Issue	Tularosa (South 506)	North of 506	Otero Mesa
Military use	High use	Moderate use with light impact	
Nonmilitary use	None	Grazing, conservation, and recreation	
Easy exit (use WSMR)	2 million adjacent acres with declining activity		
Army-wide role	Nation's Air Defense Center		

Counter Arguments?

Figure 19

Figure 19 highlights our conclusion that despite Army hesitancy in funding the renewal process, Fort Bliss and the McGregor Range play a critical military role as the nation's air defense center. The Otero Mesa and areas north of 506 are critical for minimizing deployment constraints in the large Roving Sands and for operational testing of air defense systems. They will become increasingly important as the new fiber optic network is used more extensively to simulate targets for both Roving Sands and more routine ADA training. It will allow units to train in a synthetic environment while deployed in a realistic field setting.

There are several possible explanations for the Army hesitancy to fund the EIS for the McGregor renewal. Obviously, with budgets tight there is a tendency to delay any expenditures, and the congressional requirement for the LEIS (as stipulated in 99-606) is three full years prior to the actual expiration of the renewal. Given Congress's ability to renew independently of the LEIS, there may have been substantial incentive to delay the expenditure.

Another explanation may lie with the discussion in Figures 11 and 13. There is a significant mismatch between the Army's major command structure and the national role of Fort Bliss. While TRADOC must pay the bulk of the Fort Bliss overhead costs, the units at Fort Bliss fulfill a mission that is far broader than the traditional Army school-house function.

Another possibility is related to the lack of a single Army proponent for training lands. As late as 1994 there was little awareness in the Department of Defense of the renewal process.[22] The Air Force might have been equally slow to fund its renewal efforts if failures in other land initiatives had not shocked the entire Air Force system.[23] The Navy also lags the Air Force in its efforts to renew the Bravo 20 range at Fallon Naval Air Station.

Although the precise reason may be unknowable, we conclude that the delay in Army funding was not the result of any lack of important military missions at the McGregor Range. Air defense is becoming increasingly critical and Fort Bliss is the nation's air defense center. The McGregor Range provides the physical space for fulfilling this mission.

In the next section of the briefing we examine to feasibility of moving McGregor activities to the larger and adjacent WSMR.

---

[22]In Rubenson et al, *More Than 25 Million Acres: DoD as Federal Natural and Cultural Resource Manager*, Santa Monica, CA: RAND, MR-715-OSD, 1996, we recommended that DoD establish a land-use policy planning team and that the first task of this team be to coordinate 99-606 renewals. The recommendation was not implemented and in our judgment was not viewed as a critical priority at the time.

[23]In 1995, the then Undersecretary of the Air Force, Rudy DeLeon, told the lead author of this report that the Air Force needed to ensure that the Nellis and Goldwater renewals were completed without the problems encountered in Idaho. At the time the Air Force was establishing a Ranges and Airspace Office to be a single proponent for range and airspace issues on the Air Staff.

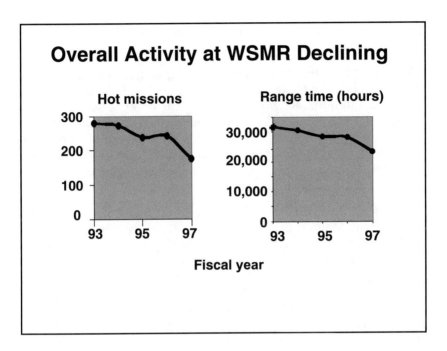

Figure 20

Figure 20 illustrates data supporting the idea that activity at WSMR is declining and that there is capacity to transfer McGregor activities to the larger base.[24] The left half of Figure 20 shows "hot missions" performed in each of five years, and the right half shows "range time." Hot items are defined as "an operation which includes a launch, drop, ejection, or other such event." Range time is the number of hours the range staff has been requested to support all missions. Since there are only 8,760 hours in a year, the numbers imply that multiple tests are typically being supported simultaneously. Both measures have dropped significantly during the last few years.

---

[24]Data taken from *Range Activity Review*, National Range Operations, White Sands Missile Range, 4th Quarter Review FY 97, RCS:EWS-NR-P-101.

Other measures of range activity discussed in the WSMR activity review show similar patterns.[25]

While WSMR retains the mission of serving as the nation's center for developmental missile testing, the aggregate data indicate an overall downward trend in activity. WSMR is "customer" financed in that there is no set amount of activity assigned and only a small budget line for the installation. Instead, individual weapon system programs bring their tests to WSMR and are charged for use of instrumentation, range support, and other services. Most fixed costs for operating the range must be covered by customer fees.

WSMR's customer financing increases the difficulty of predicting future activity. Major programs such as THAAD (Theater High Altitude Air Defense) and Patriot enhancements will continue to give WSMR important missions if they proceed on schedule. However, the declining activity, and the need to recover fixed costs from a reduced customer base, could lead to price hikes that further discourage customers from using WSMR.[26] There are already reductions in force taking place at WSMR, and long-term plans indicate the need for further personnel cuts. Given that WSMR activities are partially governed by market-like forces rather than assigned missions, all such estimates contain significant uncertainty.

Nevertheless, WSMR range officials felt confident that Patriot firings occurring on McGregor could be accommodated within WSMR's schedule. They provided RAND with an estimate of approximately $60,000 in range support costs to fire a missile. This amounts to approximately $2.4 million per year at the current firing rate. This figure corresponds to the use of only that instrumentation support

---

[25]Probably the best metric for determining whether WSMR could accommodate additional training would be utilization based on geographic use of the land and airspace. WSMR Range Control informed us that such historical data do not exist.

[26]As an example, WSMR has recently suggested a dramatic price increase in the costs it charges Fort Bliss to supply an F-16 target and for time for that vehicle to utilize WSMR airspace. WSMR has an interservice agreement with Kirtland AFB to purchase F-16 flight time. It then sells the aircraft to Bliss with additional charges for WSMR airspace use. As of two years ago, the cost to Bliss was $1,685 per airplane hour. Last year WSMR billed Bliss $2,800 per aircraft hour, though Bliss believes there is a contractual agreement for the $1,685 figure. WSMR is now proposing a fee of $4,600 per hour. Although negotiations may ultimately lower this figure, it is illustrative of the need to raise prices when activity declines in a customer-funded installation.

required for safety purposes and the use of a simple MQM-107 aerial
target.

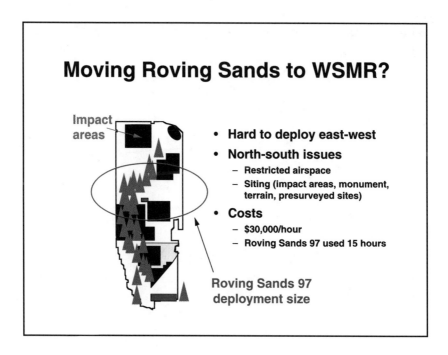

Figure 21

Figure 21 illustrates some of the major topographical and geographical features of WSMR and overlays the size of the Roving Sands 97 deployment area onto the range. The red area in the southernmost tip of WSMR is the major missile launching facility, located directly to the east of the installation's main post facilities. The yellow area is the White Sands National Monument. This world-famous tourist attraction is contained within WSMR but can be evacuated during a missile test. The black shapes are impact areas, and the brown triangular shapes represent the San Andres Mountains.

Figure 21 shows that WSMR's geometry would preclude Roving Sands 97 deployment in the current east-west formation. WSMR is narrower than McGregor, and its existing width is bisected by the San Andres Mountains. Deploying ADA units on both sides of the mountain would reduce the effectiveness of Roving Sands, since a central goal of the exercise is to test interoperability using a single picture of the airspace. Line-of-sight limitations imply that units to the west of the mountains would see a different picture than those to the east. While such situations might be consistent with some real battle scenarios, Roving Sands planners are currently interested in testing interoperability with a shared picture.

The obvious solution would be to rotate the exercise to a north-south orientation. Extensive siting and survey work would be needed to find and approve new locations for the Patriot training sites, though there is no obvious obstacle to this.[27] The more difficult challenge is to realign the airspace. The Roving Sands ADA deployment at Bliss takes advantage of the long channel of restricted airspace above WSMR. This airspace is almost never turned back to civilian authorities and runs from floor to ceiling. This allows attacking aircraft to fly diverse north-to-south profiles as they approach the air defenses located on the McGregor Range.

WSMR does not have access to low-altitude airspace directly to the east of the installation. It would be difficult to arrange use of the airspace. Such a plan would arouse public objection and work against a military goal of minimizing the need for new airspace and lands. There are also several commercial airports in the area to the east of WSMR that use this airspace for takeoffs and approaches.

The most immediate reason that planners do not consider WSMR as an alternative for Roving Sands is cost. WSMR charges approximately $30,000/hour for use of the airspace.[28] The high cost is largely due to the difference in funding structure between WSMR, Fort Bliss, and other units participating in Roving Sands. WSMR is customer funded.[29] Almost all costs for installation infrastructure including salaries, environmental studies, range maintenance, etc. are recovered from users of the range. This is a general policy for the nation's Major Range Test Facility Bases (MRTFBs). It means that funds for test activities include money to pay for range time and support.

---

[27]According to the Fort Bliss environmental office, the costs of required environmental studies for a one-square-kilometer site are approximately $70,000.

[28]Communications from Forces Command based on charges during the Roving Sands 97 exercise.

[29]Customer funding of MRTFBs is dictated by DoD directive 3200.11.

In contrast, the budgeting processes for TRADOC and Forces Command do not lend themselves to individual units paying for test facility infrastructure or even Forces Command and TRADOC infrastructure. These installations are "mission" funded. Funds for infrastructure support are separated before budgets are assigned to the operating units. Thus, while operating units are free to pay for use of WSMR, OPTEMPO budgets carry the implicit assumption that funding for ranges has already been extracted.

Figure 21 indicates that WSMR charged approximately $30,000/hour for use of the range and that Roving Sands utilized the range for approximately 15 hours during the 1997 exercise. These costs are viewed as excessively high by trainers, and Roving Sands planners seek to minimize use of WSMR. In contrast, WSMR managers must account for loss of capability to conduct test missions during the time and must charge to account for lost opportunities.

Is this cost a fair charge? One of the problems in assessing that question is the lack of WSMR tradition for billing training customers. Billing for tests is linked to the level of instrumentation support that WSMR personnel must provide. The use of WSMR by trainers merely for its vast land and airspace resource is inconsistent with WSMR's mission, its financial structure, and the type of missions it has historically supported. Consequently, WSMR bills Roving Sands as if the instrumentation that is precluded from use during the exercise is, in fact, being used.

The question regarding justification of costs nevertheless remains. The 15 hours that Roving Sands used WSMR did not imply significant instrumentation support, though it may have precluded missions that use such assets. WSMR has approximately $100 million (or about 30 to 40 percent of revenue) of annual fixed costs to maintain the installation as a center for developmental missile testing.[30] This implies the installation must recover approximately $22,000 per daylight hour of fixed costs from customers. The $30,000/hour seems

---

[30]Communication from WSMR National Range Office.

consistent with the financial needs of WSMR as it is structured to-day.[31]

It is also interesting to note that the Roving Sands 97 exercise seemed to leave both trainers and WSMR support personnel with a sense of the incompatibility. Trainers were frustrated at the prices, while WSMR found it difficult to support the small number of ground units deployed on diverse parts of the range. The WSMR work force is composed almost entirely of civilians living in El Paso, Las Cruces, and Alamogordo, and there is little infrastructure for maintaining any significant number of troops.

---

[31]As will be discussed later, the U.S. Air Force currently uses ranges at WSMR for significantly lower costs. WSMR accommodates the Air Force on an "as-available" basis and in exchange for Air Force support in airspace management.

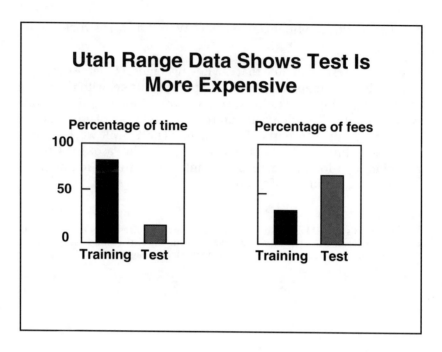

Figure 22

Figure 22 displays data from the Utah Test and Training Range (UTTR) that provide additional insight into the reasons behind the high costs of utilizing WSMR by training units.[32] UTTR is formally categorized as MRTFB and subject to DoD requirements to recover operating costs from users. However, UTTR is also the closest range for Hill Air Force Base, and as the left half of Figure 22 illustrates, most of the activity conducted at UTTR is training related. As such, UTTR has had to estimate the costs of supporting training and compare those costs with those incurred for operational testing. WSMR, in contrast, has few requests for training and has not developed a carefully derived pricing scheme for training activities.

The right side of Figure 22 shows that fees for testing constitute approximately 60 percent of UTTR revenues. The left side of Figure 22 shows that testing constitutes only a small portion of range activity.

---

[32]Data supplied by Air Combat Command.

Together these data illustrate that testing is dramatically more expensive than training. This is due to the relatively small amount of instrumentation and instrumentation support required for training.

The data support the validity of the seemingly high charges imposed by WSMR for training activities. If training truly precludes test activities, than the test range has little choice other than to charge high prices. The high costs of test, due to the extensive fixed costs, make it difficult for a customer-funded institution structured to support test to also accommodate occasional training. Internal DoD financing is thus a major obstacle to sharing the test and training resources.

This is not to say that some of these obstacles couldn't be overcome. WSMR could develop a pricing mechanism for training, and it might be financially desirable to accept low-fee training activities in the absence of test activity. However, UTTR has a significant level of training activity, while training at WSMR is currently limited to special occasions. WSMR does not yet see enough potential training business to develop new approaches.

Land and Airspace Probably Available, but...

- **North-south deployment an issue**
- **Financial/regulatory structure**
  - Test gets priority by policy
  - WSMR needs test fees
  - Training ranges mission funded
- **Long-term WSMR options?**
  - Test range consolidation, or
  - Policies encouraging co-use
- **Relocation would require:**
  - Significant changes in DoD policy
  - Environmental impact analysis

Figure 23

Figure 23 summarizes our findings on the use of WSMR for Roving Sands and more broadly as an alternative to renewal of the McGregor Range.

Although WSMR's methods of recording range usage do not lend themselves to a geographically based description of utilization, the vast spaces and declining activity make it reasonable to conclude there is sufficient territory for Roving Sands and other exercises that use the Otero Mesa and areas north of Highway 506. However, this finding itself would need to be verified with site-specific environmental studies and efforts to schedule such activities. Roving Sands would be constrained in an east-west deployment or could deploy north-south. The latter implies the need for additional civilian airspace that would be difficult to acquire given the local commercial airports to the east of WSMR. The mountain chain running north-south in WSMR would also change the type of interoperability testing that could be done.

While the physical resources of WSMR are sufficient to contemplate transfer, the financial and internal DoD regulatory systems for Bliss and WSMR currently prohibit this consideration. Test customers get priority by DoD policy, and WSMR needs test customers to cover the fixed costs associated with its core mission.[33] As seen by the Roving Sands charges, attempting to recover fixed costs from trainers is essentially prohibitive.

Some have suggested that the end of the Cold War and declining activity at WSMR should allow the return of some military land to the public domain. For the Bliss/WSMR complex to represent such an opportunity, changes in policy consistent with a congressional Base Realignment and Closure (BRAC) process, rather than the 99-606 process, seem more appropriate. Despite declining activity, WSMR remains the nation's center for developmental missile testing. Declining business implies that the prices it must charge to recover fixed costs could become unrealistic. Congress and the DoD would then want to examine a range of options, such as bringing in additional test business through a consolidation of the MRTFB structure. The idea of transferring a portion of the Bliss mission to WSMR is only one option.

Contemplating a shift of some Bliss activities to WSMR would take significant DoD policy changes and the development of new operating practices. Tests tend to be scheduled at precise times and are difficult to change. Training occurs over longer periods and is subject to far more unpredictable events. As noted above, the financial structures are also incompatible. Nevertheless, the long-run need for more efficient use of land may necessitate such a blending. DoD might consider an experimental and evolutionary process that seeks to overcome financial obstacles and helps develop new operating practices so that eventual consolidation might be a more realistic policy option.

_____

[33]WSMR has a three-tiered priority system: (1) research, development, testing, experimentation, guided-missile firings, and high-energy laser operations, (2) non-research guided-missile firings development, testing, and experimentation, and (3) other missions including training. See *White Sands Missile Range Range-Wide Environmental Impact Statement,* January 1998.

Finally, we note that even if internal DoD financial and procedural policies could be overcome, extensive environmental impact analysis would be required to change the location of activities. This can involve substantial financial resources and long periods of time to complete.

None of these factors implies that a transfer of Otero and north of 506 activities to WSMR would be impossible. However, the 99-606 process is not creating the political will that could produce the necessary policy changes. As such, it is impossible for those fulfilling the renewal effort to consider this option. The initiative would need to come from senior leaders of DoD or Congress.

Arguments for Limiting Renewal?

Issue	Tularosa (South 506)	North of 506	Otero Mesa
Military use	High use	Moderate use with light impact	
Nonmilitary use	None	Grazing, conservation, and recreation	
Easy exit (use WSMR)	None	Tactical penalties, Option not available	
Army-wide role	Nation's Air Defense Center		

Counter Arguments?

Figure 24

Figure 24 incorporates the findings of the previous section in the policy matrix. There appear to be sufficient natural resources (land and airspace) to accommodate McGregor activities at WSMR. However, the significant internal DoD policy and procedural issues, the loss of optimal geometry for some exercises, and the necessary environmental analysis essentially preclude this option at this time. It seems doubtful it could be prepared in time for the 2001 renewal. It also seems imprudent to view WSMR as a permanent site for McGregor activities in the absence of a DoD-wide assessment of test ranges. One option for WSMR is to gain increased utilization through the consolidation of other test activities on the range.

We have chosen to color the box green for Otero and north of 506 because this option is not available to planners without leadership from Congress or senior DoD policymakers; hence it supports the case for renewal. It would also require extensive planning and environmental analysis.

The matrix in Figure 24 indicates that it is difficult to build an "airtight" case for the complete renewal of the range without considering the proposed bombing range on Otero Mesa. However, even without the bombing range there are important military uses of the entire range that would be difficult to reproduce elsewhere. A key consideration therefore becomes the extent to which military activities interfere with nonmilitary uses. We examine this question in the next section of the briefing.

Nonmilitary Uses

- **Cattle grazing**
  - ~ 2,500 head
- **Recreation / hunting**
  - ~ 1,500 permits per year
- **Habitat**
  - Grasslands
  - Potential habitat alpomado falcon
  - Cattle grazing a source of damage?
  - Damage from today's military mission?

Figure 25

USES

The nonmilitary uses of the Otero Mesa and the region north of 506 are summarized in Figure 25.

The use of the McGregor Range for grazing is important for ranchers in the region who use the range, particularly as a reserve in times of drought. Although numbers vary annually, about 2,500 cattle graze Otero Mesa and areas north of 506. The extensive water pipeline and Army water rights allow grazing leases to be sold at prices many times higher than that mandated on public lands. This is particularly true in drought years, when ranchers are anxious to bring cattle to the range's water resources. For example, in 1994 near the end of a several-year drought, grazing leases were $13–$14/AUM (animal unit monthly). According to the draft LEIS, average prices in some years have been as high as $16.75/AUM. This compares with the $1.35/AUM on leases subject to the Taylor Grazing Act.

Another nonmilitary use of the range involves various forms of recreation such as hunting, hiking, or merely enjoying the solitude. About 1,500 permits each year are granted for access to all of Fort Bliss.

The Otero Mesa and Sacramento Foothills are often cited for their importance as habitat. This nonuse value obviously enhances some of the recreational uses, but is perceived by many as a value by itself. Laws such as the Endangered Species Act and other statutes aimed at protecting habitat and wildlife provide evidence of society's acceptance of these values.

SOURCES OF DEGRADATION?

Does today's military mission degrade the habitat? Or is ecological damage attributable to cattle grazing? Use of rangelands by cattle tends to be nonuniform, and the use of the Otero Mesa on the McGregor Range is no different. Cattle tend to stay near the water troughs, so the immediate area around the troughs on McGregor is often reduced to bare ground from cattle traffic, resting cattle, and accumulated feces. However, this damage is generally limited to the immediate area surrounding the water supply and is not representative of the remainder of the mesa.

While the McGregor Range might be characterized by some definitions as overgrazed, there are several reasons to believe the condition is still relatively good. One indicator is the presence of a sensitive species such as the chestnut-collared longspur *(Calcarius ornatus)*, a bird known to be in decline owing to the loss of its preferred prairie habitat. The mesa represents a wintering ground for this species. It isn't clear to what extent the decline is due to loss of habitat on its breeding grounds versus its wintering grounds. Nevertheless, its presence on the mesa speaks to the condition of that portion of the range used for grazing cattle, other nonmilitary activities, and military activities.

Another indicator of the condition of the Otero Mesa is the lack of significant invasion by nonnative annual grasses. Native, warm season, perennial grasses like blue grama *(Bouteloua gracilis)*, black grama *(Bouteloua eriopoda)*, sideoats grama *(Bouteloua curtipendula)*, and threeawns *(Aristida species)* still predominate on the mesa. These shortgrass prairie constituents, along with soaptree yucca *(Yucca elata)* and banana yucca *(Yucca baccata)* and tree cholla (*Opuntia imbricata),* are the key characteristic components of native Chihuahuan desert grasslands. Yet a third, related indicator is that the mesa grasslands have not been converted into shrublands to any extent. Shrublands not only represent a loss of native grassland from a natural resources perspective, but also represent a less productive forage for cattle. While mesquite and opuntia cactus have become abundant on overgrazed rangeland in many portions of the Chihuahuan desert, such shrubs are still relatively sparse on Otero Mesa.

One species that occupies Chihuahuan desert grassland is the aplomado falcon *(Falco femoralis),* a species whose range extends south into the neotropics and which still breeds (in declining numbers) in Mexico. There are three distinct subspecies of aplomado falcon, one of which, *Falco femoralis septentrionalis,* is on the federal endangered species list and is therefore a subject of concern on the McGregor Range. This falcon uses abandoned nests of ravens or hawks atop a low mesquite or yucca from 8 to 25 feet above the ground. However, most of the soaptree yucca, the tallest woody vegetation on Otero Mesa, do not appear to exceed 8 feet there.

Although the last breeding record for a truly wild population of this falcon in the United States is in 1940, the Army and the U.S. Fish and Wildlife Service seek to manage potential habitat as though it contained aplomado falcons. Since the grasslands of the Otero Mesa provide potential habitat for this species and sufficient size to accommodate its large nesting territory and foraging requirements (mainly small rodents, lizards, and snakes), there is pressure from environmental organizations to manage the mesa for the aplomado falcon. However, sightings of this bird in the United States are extremely rare (it has been reintroduced to south coastal Texas at Laguna Atascosa National Wildlife Refuge). The only sighting in recent decades on the mesa was an immature bird. (Immature birds are known to wander more than adults from the species' normal range.) At any rate, current Army ADA training and testing activities on the mesa do not appear to conflict with management of this potential habitat for this federally listed falcon.

It is not a coincidence that the last record of aplomado falcon from the lower Sulphur Springs Valley was November 1939, a year after the disappearance of the last black-tailed prairie dogs in southeastern Arizona. Yet on Otero Mesa, there are still prairie dog towns to be found with viable populations that are monitored by the natural resource managers on Fort Bliss. This is another indication that the current military and nonmilitary uses of this portion of the range are compatible with protection of such natural resources. The major damage to the mesa appears to come from cattle grazing around water supply points.

The remaining question is the extent to which the current military mission limits access to the range. We will discuss this in the context of overall BLM management goals in the following chart.

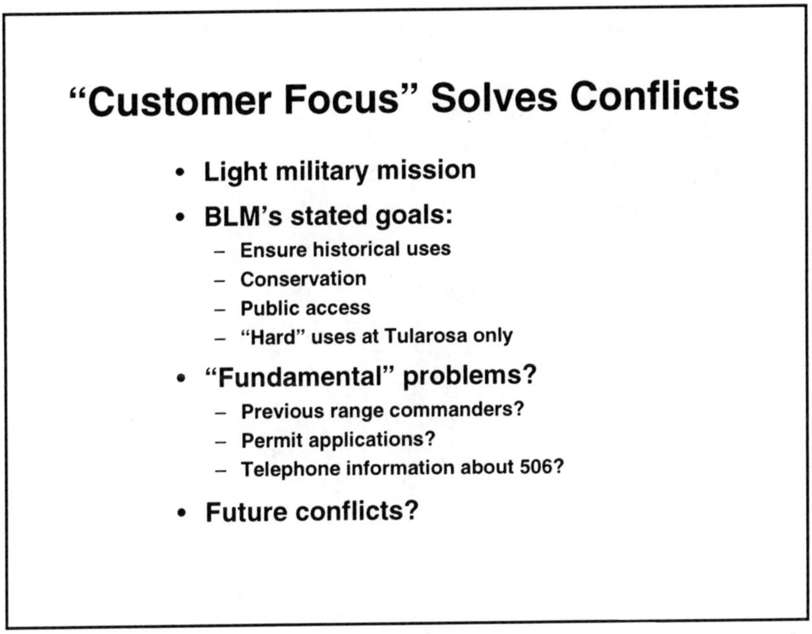

"Customer Focus" Solves Conflicts

- **Light military mission**
- **BLM's stated goals:**
  - Ensure historical uses
  - Conservation
  - Public access
  - "Hard" uses at Tularosa only
- **"Fundamental" problems?**
  - Previous range commanders?
  - Permit applications?
  - Telephone information about 506?
- **Future conflicts?**

Figure 26

Figure 26 expands our conclusion that today's military mission has not had a significant adverse effect on habitat to a more general conclusion regarding nonmilitary use. It also summarizes Bureau of Land Management goals as stated at public scoping meetings for the LEIS. BLM manages many nonmilitary uses on the McGregor Range; subject to control of access by the Army. The major problem in this relationship has been the Army's lack of tradition for serving external constituents, not the resource constraints arising from dual use.

As indicated in Figure 8, there is minimal scheduled ground military activity that has a direct impact on land on Otero Mesa or north of 506. Most use is for low-altitude overflight and safety fans. The limited military ground mission, such as deployment of Roving Sands, involves road movement of nontracked vehicles, with off-road maneuver being confined to presurveyed and carefully monitored one-square-kilometer boxes. The army spends roughly $70,000 for the environmental studies required to site such locations.[34]

At the public scoping meetings[35] the BLM stated that its McGregor management goals are to maintain historical uses (primarily grazing), protect natural resources, ensure public access, and confine "hard" military use to the Tularosa Basin. "Hard" use is defined as military missions with direct ground impact.

The current military impacts and BLM goals do not seem incompatible. Nevertheless, comments from scoping meetings and other communications indicate disagreements driven by issues such as those highlighted under the third bullet of Figure 26. None involve natural resource constraints that would preclude the existing levels of joint military and nonmilitary use. The resource is sufficiently large, the impacts sufficiently small, and the levels of use sufficiently low, to accommodate both uses. Instead, conflict and disagreement appear to be related to managerial procedures and processes such as nonresponsive range commanders, slow action on access permits, lack of easy access to telephone information on the status of Highway 506 (which the Army can close for safety purposes), and other issues related to ease of access. The following examples are illustrative of how the current management system has made the land more difficult to manage for nonmilitary uses:

- The Army conducted non-time-urgent environmental surveys during optimal hunting periods, thereby limiting hunters' access to the range.

- The BLM's ability to initiate the access permit process has been limited to residents of Las Cruces. Residents of other areas have had to drive to Fort Bliss to initiate the process.[36]

---

[34]Based on communication from the Fort Bliss environmental office and averaging of the estimated costs for a set of sites currently being developed.

[35]Held November 18, 19, and 20, 1997, in Alamogordo, Las Cruces, and El Paso. See *McGregor Range, New Mexico Land Withdrawal Renewal, Scoping Summary,* February 1998, for complete transcripts and follow-up written public comment.

[36]Although complaints about the access permit process at the scoping meetings were clear, the problems are somewhat puzzling to us. Permits are granted on an annual basis and can be acquired at the McGregor Range control office any time during the year. The office is located about 20 miles north of El Paso. Three to four times a year there is an opportunity to get an annual permit in Alamogordo, twice a year at

- The Army divides McGregor by zones of military use and controls access by this zoning system (the zones are displayed in Figure 28). The result has been that entire zones have been off-limits even when only very small surveys involving one or two people were involved.

- Recent BLM efforts to build a small storage shed were initially rejected without discussion (though later addressed).

There are undoubtedly more examples of managerial tension in the Army-BLM relationship. However, the problems seem to reflect the Army's unfamiliarity with dealing with outside constituents rather than resource constraints or any inherent incompatibility between the Army and BLM missions there. Our interviews did not reveal any fundamental conflicts at the existing levels of use.

In recent months the Army has made several changes in its approach toward facilitating outside use. The Army is discussing the storage shed with BLM. The Army is attempting to go to additional sites, such as Walmart stores, to offer access permits. More significant is the extensive effort made to notify the public and seek input for the renewal process itself. A broader philosophy of viewing the public as a constituent should eliminate most of the tension associated with nonmilitary use of the range.

We should also note that the current arrangement provides two advantages for the BLM's grazing program. First, the withdrawal allows lease prices to be established by auction rather than the regulated prices under the Taylor Grazing Act.[37] This produces roughly an order of magnitude increase in revenue, allowing BLM to pay the salaries of several range managers who maintain and operate the water pipeline system. The second advantage is that the cattle in the BLM grazing program use water fed through a pipeline system on McGregor. The Army holds water rights, and Army funds were used

---

Holloman AFB, and twice a month at the main cantonment area in Fort Bliss. Permit holders must telephone range control on the day they hope to use the range to ensure that it is open.

[37]See Fowler, Torell, and Gallacher, "Competitive Pricing for the McGregor Range: Implications for Federal Grazing Fees," *Journal of Range Management*, Vol. 47, March 1994, pp. 155–158.

to acquire major portions of the pipeline.[38] If the land reverted to the BLM, these advantages would be lost unless Congress explicitly chose to maintain them in new legislation.

More difficult issues involve future uses and whether the current arrangement represents a compromise that is unacceptable to both military and nonmilitary users. Several written comments submitted to Fort Bliss suggest the potential for significantly increased recreational use if access permits were not required.[39] Several at the public hearings (and in written comments submitted to Fort Bliss) spoke to a need to transfer parts of the Otero Mesa and Sacramento Foothills into Wilderness or National Conservation Areas in order to ensure preservation of ecological values. However, the letters cited few incidents of ecological damage caused by ongoing military activities, and we assume they are oriented toward ensuring against future damage.

Until the bombing range proposal, the military made few efforts to use the Otero Mesa in a more intensive manner. Some Fort Bliss personnel feel that there would be military benefits to such use (see the discussion in Figure 29) but that a tradition of limiting use on Otero Mesa and north of 506 has precluded consideration of such options. Currently the assumed need to maintain dual use has eliminated many suggestions.

The Bureau of Land Management believes that the proposal for a new Air Force bombing range on Otero Mesa has changed the accepted balance. This proposal will be discussed later in our sensitivity analyses.

---

[38]According to Fowler et al., the availability of water and the sunk capital costs make the auctioned price comparable to Taylor grazing prices from the perspective of a customer. Since BLM pays no finance charge on the capital investment, it uses the fees to pay salaries. In comparable circumstances without the pipeline, the grazing price would be lower (Taylor Act) but ranchers would need to pay more than the difference to secure water.

[39]See *McGregor Range, New Mexico Land Withdrawal Renewal, Scoping Summary,* February 1998.

There Are Few Arguments for Limiting the Renewal

Issue	Tularosa (South 506)	North of 506	Otero Mesa
Military use	High use	Moderate use with light impact	
Nonmilitary use	None	Conflicts resolved with "customer focus"	
Easy exit (use WSMR)	None	Tactical penalties, Option not available	
Army-wide role	Nation's Air Defense Center		

Alternative divisions, bombing range?

Figure 27

Figure 27 incorporates our findings into the policy matrix. As stated above, we find no fundamental conflict between the current military and nonmilitary uses of the Otero Mesa and the area north of 506. *The light impact of the current military mission makes the Otero Mesa/north of 506 area suitable for dual use, especially since the nonmilitary use has been limited to the grazing of approximately 2,500 cattle and distribution of 1,500 access permits per year.* The Army's ongoing efforts to gain greater sensitivity to the competing priorities should reduce many of the problems of the past.

This compatibility would seem to override the strongest arguments for limiting the renewal. Relatively minor changes in the Army/BLM working arrangement could eliminate most of the existing conflicts on Otero and north of 506.

In the next section we consider sensitivity analyses involving different divisions of the range and the proposed U.S. Air Force bombing range on Otero Mesa.

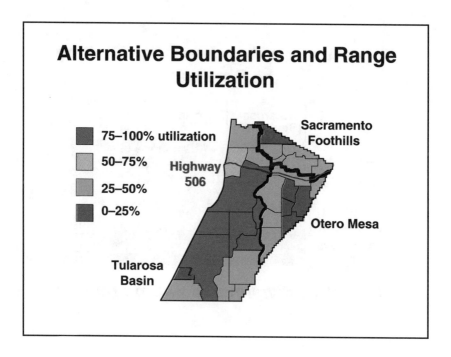

Figure 28

Figure 28 shows range utilization within each of the maneuver areas used to divide the range.[40]   Usage is divided into four utilization groups, as indicated in the legend. The thick black line highlights a division of the range that corresponds to geological boundaries rather than nonmilitary use. In this division, the Tularosa Basin extends to the northwestern corner of the range and part of the Otero Mesa extends across Highway 506. The northernmost area is the Sacramento Foothills.

The figure illustrates that a geological division doesn't significantly change the preceding considerations. One major distinction is the moderately high-use area north of Highway 506 on the western edge of the range. The inclusion of this piece in the Tularosa makes the

---

[40]These data were adapted from the Draft Environmental Impact Statement for the McGregor Range as presented in Table 2.1-2 of that document. Data in Figures 8 and 9 were extracted from our own analysis of the 1996 range utilization data.

Sacramento Foothills a lower-use region than the entire region north of Highway 506.

In the next section we will consider an option for increased usage on the Otero Mesa. Under this situation the Sacramento Foothills become the single low-usage area of the range. A willingness to pay the $60,000 per Patriot firing at WSMR would further reduce the usage of the Sacramento Foothills. Thus the Sacramento Foothills might be the most easily separable part of the military range, should Congress feel a necessity to return part of the land to the public domain. Almost all of the very small amount of recreational use occurring on McGregor takes place in the Sacramento Foothills. The low use is probably due to the remoteness of the area and the availability of many public lands in this area of New Mexico.

Future Uses?

	Tularosa	Otero Mesa	North of 506
In process		• Bombing range • Patriot sites	
Informal planning	• MOUT	• TEWTs	• Target launch facility
Speculative	• Guard heavy training • Helicopter gunnery		

Figure 29

Future military uses of McGregor vary from speculative ideas to approved plans that have gone through extensive environmental review and received numerous public comments. Figure 29 highlights potential future military uses of the McGregor Range.

The Tularosa Basin is a vast open area ideal for planners to postulate military training activities. One earlier plan that was never implemented is an Apache helicopter gunnery range. There have also been recurring discussions about the possibility of opening up maneuver corridors within the basin. The 3rd Armored Cavalry went as far as asking the environmental office for preliminary assessments, but ultimately balked at the costs of the required EIS. This idea has been revived along with discussions about making Fort Bliss a center for National Guard training. The costs for any ordnance removal would also need to be examined, as the basin has been used for firing for more than four decades. No funds have been allocated to pursue this idea.

The National Guard has not considered Otero Mesa for the national training site even though it seems ideal for conducting military maneuvers. As discussed in Figure 26, the informal understanding on dual use has discouraged Army planners from considering new ideas for using Otero Mesa.

One exception is the continuing implementation of one-square-kilometer training boxes for Patriot units. About 15 sites already exist for Roving Sands and about 25 more are in the planning stages for use in regular ADA training.[41] However, there will be no off-road maneuver of tracked vehicles. More speculative are discussions on Training Exercises Without Troops (TEWTs) on the Otero Mesa. These are exercises oriented toward officers in which plans and movements are implemented without tactical vehicles. The notion is to conduct planning for a partially real exercise. Also being considered is a Military Operations on Urbanized Terrain (MOUT) complex on the Tularosa Basin consisting of 32 buildings.

The growing priority of the missile defense component of air defense has created the need for longer-range targets. At this time Fort Bliss is considering a new launch site on the very northwestern corner of the McGregor Range, in the area north of 506. Targets intended to simulate the flight characteristics of a SCUD missile would be launched from this area. This facility would be built in the portion of the Tularosa Basin north of Highway 506 and not in the Sacramento Foothills.

None of these uses individually would fundamentally alter the policy matrix discussed in Figure 27. But some feel that even these low-impact and limited activities mark a gradual trend toward more intensive use and eventual elimination of the nonmilitary activities. As noted above, several of the activities mentioned in Figure 29 are speculative.

The one activity that could alter the relationship between military and nonmilitary use is the U.S. Air Force proposal to build a bombing

---

[41]Forces Command Environmental office has begun monitoring the one-square-kilometer plots. The two years of data show degradation after the Roving Sands exercise and a return of vegetation the following year. There are, however, not enough years to draw firm conclusions.

range on the Otero Mesa. This proposal and its implications will be
discussed in the following section.

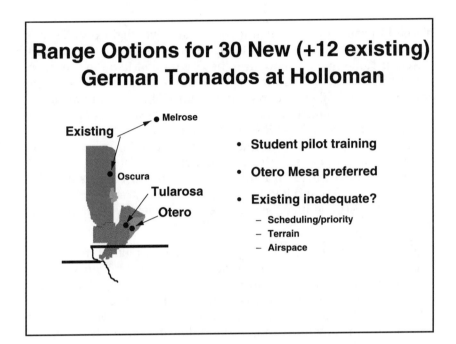

Range Options for 30 New (+12 existing) German Tornados at Holloman

- Student pilot training
- Otero Mesa preferred
- Existing inadequate?
  - Scheduling/priority
  - Terrain
  - Airspace

Figure 30

Figure 30 provides an overview of the issues that led the U.S. Air Force to propose construction of a bombing range on Otero Mesa. The proposal has gone through public hearings and produced a Final Environmental Impact Statement (FEIS). This April the U.S. Air Force issued a Record of Decision (ROD) to proceed with the range. Since no new land must be withdrawn, the range can be built without new legislation. At this time we expect the range to be built unless there are successful lawsuits.

The origin of the request is the joint decision by the U.S. and German governments to bed down 30 additional Tornado aircraft at Holloman Air Force Base. Twelve other Tornados are already in place.[42] The new aircraft will be used primarily for providing new pilots the

---

[42]For a brief history of German air force involvement at Holloman, see Air Combat Command, *Proposed Expansion of German Air Force Operations at Holloman AFB, New Mexico, Operational Impact Statement*, October 6, 1997.

Tornado Basic Course, which involves low-level flight profiles and use of terrain following radar (TFR). Both German and U.S. aircraft based at Holloman currently use the Oscura and Red Rio ranges at WSMR.[43] Access to the range is provided on an "as available" basis for only a minimal charge in exchange for Holloman support in air traffic control management. *This is a relationship fundamentally different from that of Fort Bliss with WSMR.* Bliss does not have an agreement for "as available" use and must pay for every activity. However, even under the Air Force's current arrangement, test missions take priority over Air Force training. The Air Force is the only user of the Oscura and Red Rio ranges, but safety fan requirements for missile tests conducted elsewhere can interfere with training activities.

Figure 30 also highlights the arguments suggesting that the current arrangement is inadequate and that there is a need for a new range. The new Tornados represent a net growth of 2,600 sorties out of Holloman, and the Air Force is seeking additional range capacity. The German air force also requires a special set of range conditions for student pilots while using terrain following radar. It prefers 10 nautical miles of flat terrain, varying no more than 200 feet in elevation, on approach to the target. It also prefers few elevation changes behind the target so that pilots remain in the TFR mode. Sharp elevation changes will force pilots to exit the TFR mode and use visual.

The Air Force EIS considered several alternatives. The existing ranges were considered to be viable but not optimal due to the factors highlighted in Figure 30. A range in the Tularosa Basin had two problems: cost and terrain. Due to the unknown levels of ordnance contamination, cleanup would have to precede construction. This elevated the estimated costs to $19 million as opposed to $4.1 million for a new range on the western portion of Otero.[44] More significantly, the location of the Tularosa range near the Otero Mesa bluffs

---

[43]Data from the WSMR range control office indicate that the Air Force was able to access these ranges for about 4,700 hours during FY 97. The ranges are open 11 hours per day, and hence there are a possible 8,030 operating hours per year for the two ranges. As noted above, it is the use of the ranges for safety fans and other airspace uses that restrict Air Force use rather than other users of the range.

[44]Data taken from the U.S. Air Force's Environmental Impact Statement, *Proposed Expansion of German Air Force Operations at Holloman AFB,* New Mexico, 1998.

reduced the number of approach angles and widened the return profile after a target pass. This led to a 40 percent reduction in target passes per sortie.[45] Finally, the Melrose Range, located near Cannon Air Force Base 180 miles to the north, was also considered. Although Melrose has both adequate airspace and appropriate terrain, the Air Force argued that the distance from Holloman implies 90 percent fewer opportunities for target passes for each sortie.

This left a new range on the western portion of Otero Mesa as the preferred alternative. Had this location been in the Tularosa Basin, there would have been little objection. But the location in the Otero Mesa led an unusual coalition of environmentalists, ranchers, and hunters to voice their opposition at public hearings held by the Air Force.[46]

Several questions emerge from the proposal. One is the extent to which a new bombing range is incompatible with nonmilitary uses. The location, at a minimum, violates the BLM goal of limiting "hard" military uses to the Tularosa Basin. However, the impact area itself is limited to a 2 × 4 mile box on the western edge of the mesa. A 12 × 15 mile safety fan will be designated around the box, with access allowed prior to 7:00 A.M. daily, after 4:30 on Friday afternoons, and on weekends.

---

[45]Communication from the German air force. The Operational Impact Statement reported a 20 percent reduction.

[46]See *McGregor Range Issues,* a compilation of information put together by the Las Cruces District Office of the BLM, January 1997, for several newspaper articles describing the opposition to the range.

RAND Arroyo Center Conducted Independent Review

- **Otero Mesa preferred military alternative**
- **Existing range option:**
  - **Not fully described in EIS**
  - **Falls short of Otero, how much?**
- **Uncertainty due to physical and institutional obstacles**
  - **Schedule**
  - **Terrain**
  - **Airspace**
  - **Military requirements**

Figure 31a

As part of our analysis of the McGregor withdrawal, we conducted a review of the Air Force need for a bombing range on the Otero Mesa. Our findings are summarized in Figure 31a. Although numerous alternatives were studied in the EIS, the politically important choice was between a new range on Otero Mesa and finding a way to use existing ranges. In our judgment, the Environmental Impact Statement did not contain enough information to fully evaluate the use of existing ranges. However, our efforts to gather supplementary information lead us to concur with the judgment that a new range on Otero Mesa is the preferred military alternative. This would be the case even if key institutional obstacles on WSMR could be removed.[47]

---

[47]We qualify this remark in that we are referring to a comparison of the new range on the Otero Mesa with use of the existing ranges on WSMR and the Melrose Range at Cannon Air Force Base. It is possible that an ideal site could be found elsewhere on WSMR, and if institutional obstacles (such as competing uses) were removed, this might become the preferred alternative. This option was not pursued in the EIS and we did not investigate it either.

Uncertainty about those obstacles makes it difficult to determine how closely the existing range option could be made to satisfy military needs.

This is an important consideration, since the new range is inconsistent with at least one of BLM's goals for nonmilitary use: confining hard military use to the Tularosa Basin. Whether Congress intended such goals to be maintained is not clear; hence the new bombing range may lead Congress to be more explicit about nonmilitary use. The uncertainty is also important because it further highlights the institutional barriers for training use of WSMR. This too is important, should Congress seek to return some military land to the public domain.

SCHEDULE

As mentioned above, the U.S. Air Force has an agreement with WSMR for low-cost, as-available use of bombing ranges on WSMR in exchange for support in air traffic control management. WSMR retains the authority to cancel training activities when they conflict with test. One issue has been the impact of cancellations on training. According to WSMR G-3,[48] Holloman aircraft used approximately 2,000 hours of range time on Oscura in FY 97 and about 1,800 of this occurred in blocks of six or more hours. WSMR reported that no Holloman mission was cancelled due to a conflict with test activities in FY 97 once the mission was placed on the 72-hour activity plan.

Holloman Air Force Base tracks sorties completed and provided us with a different set of data for FY 97. These data indicate that 941 out of 10,346 sorties had been canceled within one day of schedule and 43 were canceled with the pilots ready to go. One Holloman scheduling official suggested that WSMR's report of zero cancellations might be the result of counting cancellations only when a block time is eliminated, rather than just reduced. In the latter case, Holloman would still lose sorties. WSMR was unable to confirm or refute this possibility, as a recent reduction-in-force had led to a loss of institutional memory regarding the exact meaning of the data.

---

[48]WSMR G-3 is a new office with responsibilities for scheduling the range and keeping track of its use. It is composed of individuals formerly with the National Range Office.

Assuming Holloman's data are correct, we can still conclude that Holloman has made extensive "as-available" use of Oscura, despite the institutional problems of accessing a test range. We are unaware of efforts to project overall WSMR activity and determine if the amount of "as-available" time will grow. WSMR's customer-funded fee structure prevents any "guarantee" of range time to accommodate the additional 2,600 sorties. However, the Otero Range has an estimated $4.1 million upfront cost, with unspecified annual upkeep charges. Since WSMR has in effect no cost-base pricing mechanism for training missions, it is difficult to assess the amount of range time these funds could purchase. Given our other data on WSMR costs, however, it seems unlikely that this would have been sufficient to satisfy Air Force needs.[49]

TERRAIN

The primary impetus for the new range is the addition of 30 new Tornado aircraft at Holloman for training German air force student pilots. One of the primary requirements is for these pilots to fly very low altitude target approaches using terrain following radar (TFR). Correspondingly, the German air force is seeking a bombing range with an extremely flat approach to a target and a flat backdrop. An elevated backdrop might force the pilot to pull out of the TFR mode into a visual flight mode, thereby precluding options to make multiple passes without exiting the TFR. Correspondingly, the EIS stated that the terrain requirements were less than a 200-foot elevation change in the 10-nautical-mile approach to the target, and no more than a 1,000-foot change anywhere 6 nautical miles in front of the aircraft (along the direction of the velocity of the aircraft, as the radar

---

[49]Cancellations on Oscura and Red Rio are due to missile safety fans filling the ranges' airspace rather than competing users on the ranges. Since this implies that a significant fraction of WSMR's airspace is in use, a first estimate of $30,000/hour might be used for purchase of the range time. This was the figure charged to Roving Sands for use of the entire WSMR airspace and is approximately the fee WSMR must collect to cover fixed operating costs. $4.1 million would only provide 130 hours of range time. Even if this is an underestimate by a factor of ten, the money seems insufficient to satisfy Air Force needs over several years. The Air Force plans to use the new range on Otero approximately 2,500 hours/year in addition to maintaining use on Oscura and Red Rio.

looks straight ahead) anywhere along the circular path necessary for making repeated TFR passes.

Figures 31b and 31c illustrate these paths for the Oscura Range on WSMR and the proposed range on Otero Mesa. Figure 31b highlights the primary advantages of the Oscura Range, while Figure 31c provides the most compelling rationale for seeking a bombing range on the Otero Mesa. Figure 31b shows that the Oscura Range has an extremely flat approach but does not meet the requirements specified in the EIS; it rises approximately 300 feet in the 10-nautical-mile approach. Still, this is probably superior to the Otero Mesa approach, which technically meets the 200-foot requirement but only because of a significant dip in the terrain. The Otero Mesa approach rises almost 500 feet in the last 5 nautical miles.

Figure 31c describes the turnaround, which is the most significant advantage of the Otero Mesa bombing range. The mesa range's flat

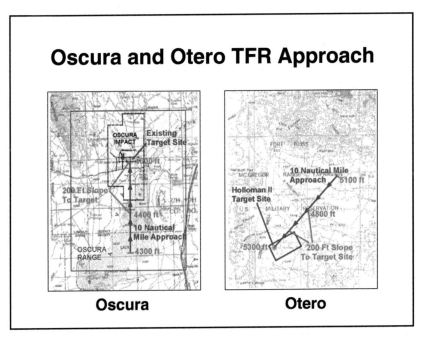

Figure 31b

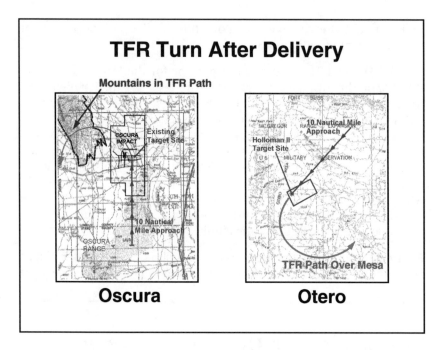

Figure 31c

turnout would allow aircraft in a TFR mode to loop around and complete another TFR pass of the target without exiting the TFR mode. German trainers feel this is an important range characteristic, since they want the ability to "drill" student pilots one technique at a time. As shown in Figure 31c, there are significant mountains to the northwest of the Oscura target site, and civilian airspace to the east precludes an eastward turn. To complete a circle back to the target, German pilots would need to exit the TFR mode.

AIRSPACE

Low-altitude airspace must be assigned along the approach and return. German air force trainers were concerned that the current airspace divisions in WSMR, and particularly those leading up to the Oscura target, were only marginally adequate for the student pilots who would be flying the TFR target passes. The EIS stated a requirement for a minimum of 12 nautical miles of airspace in front of the target. The airspace boundary illustrated in Figure 31c falls just short of this figure. South of this boundary is the Salinas Corridor, which is used to move other military traffic across the range. Our conversations with WSMR air traffic control personnel indicated that it should be possible to assign the Salinas Corridor to the Air Force for significant periods. But until efforts to do so are negotiated and demonstrated, this remains a significant uncertainty.

MILITARY REQUIREMENTS

The EIS bases desired range characteristics on sortie requirements. An ideal sortie for a German student training would optimally consist of about 10 target passes. These might be visual, TFR, or "pop-up."[50] However, the TFR mode presents the most demanding terrain requirements and hence determines terrain requirements for the range. As noted above, German trainers would like the ability to conduct sorties containing consecutive TFR passes without having to exit the TFR mode. They would also like the capability to make these passes after conducting 20 minutes of low-altitude navigation over Military Training Routes (MTRs). The combination of navigation and target passes is seen as replicating realistic sorties.

This latter requirement is critical because it eliminates the Melrose Range as an alternative for TFR passes. Melrose has adequately flat terrain but is not directly accessible for low-altitude navigation from Holloman. Aircraft fuel capacity allows time for only one or two passes of Melrose by the time pilots from Holloman reach the MTRs. This is in contrast to ten passes for Otero. Without a requirement for combining low-altitude navigation and target passes, pilots could fly directly to Melrose, complete at least six target passes, and return at high altitude.

---

[50]Pop-up bomb deliveries consist of an initial approach to the target area at low level followed by a rapid climb at a previously computed distance from the target. At a predetermined altitude the aircraft is rolled upside-down and the nose pulled down to between 10 and 30 degrees nose low. The target is then aligned with the Heads Up Display (HUD), the aircraft stabilized in one-G wings-level flight, and the bombs released. The pilot then climbs away to avoid the ground and bomb fragmentation pattern. During this maneuver the aircraft rapidly gains, and then rapidly loses, several thousand feet.

As noted above, the EIS states requirements in sorties and not target passes. We had hoped the German air force training syllabus would provide detailed requirements on the types of target passes needed, how individual sorties use the three different types of passes, and the feasibility of separating the TFR passes and conducting them on Melrose. Visual sorties could then be flown on Red Rio or Oscura. We were denied a copy of the document but allowed to view it at one sitting. Our impression was that the syllabus did not specify these needs in great detail and left significant discretion to the on-site instructors. In other words, the syllabus did not appear to state whether all sorties could be pure TFR or were required to mix TFR, visual, and pop-up.

CONCLUSION

The combination of terrain conditions and proximity to Holloman gives military trainers significant flexibility and makes the new range on Otero Mesa the preferred military alternative. However, by flying most TFR passes at Melrose, modifying some military requirements, determining if "as-available" time might grow, most of the training requirements could be fulfilled. In our judgment, the traditional obstacles to using WSMR for training led Holloman and the Air Force away from enthusiastic pursuit of this option.

Policy Matrix with Bombing Range

Issue	Tularosa (South 506)	North of 506	Otero Mesa
Military use	High use	Extensive use	
Nonmilitary use	None	Skepticism about long-term viability	
Easy exit (use WSMR)	None	Tactical penalties, Option not available	
Army-wide role	Nation's Air Defense Center		

Figure 32

Figure 32 recasts the policy matrix using the assumption that the Otero Mesa bombing range will be built. As indicated by the green color, the bombing range will mean extensive military use for (the western part of) Otero Mesa and should eliminate discussion about the level of military activity.

A more interesting question regards the nonmilitary uses on the mesa. Technically the range will comprise less than 2 percent of the nonmilitary use area and should not have a significant impact on cattle grazing. However, BLM officials believe that the bombing range demonstrates the military's ability to undertake actions that could reduce nonmilitary uses. They also point to the impact of aircraft noise on the several hundred recreational users of the range. We should, however, reemphasize that the nonmilitary uses of the range are little more than 2,500 grazing cows and 1,500 recreational permits per year.

The appendix includes a review of the legislative history of Public Law 99-606. It appears that Congress recognized the potential for military projects to erode nonmilitary uses and, while not encouraging it, gave the military that authority.

Conclusions

- **McGregor Range plays a critical role in Fort Bliss's national air defense mission**

- **Otero Mesa, north of 506**
  - Moderately used, but important military value
  - Nonmilitary uses

- **Dual use practical with improved customer focus**

- **Otero Mesa bombing range changes conclusions**
  - More significant military use
  - Questions about nonmilitary use

Figure 33

Figure 33 summarizes the findings presented in Figures 27 and 32. Fort Bliss has a critical national military mission and the McGregor Range plays a critical role in fulfilling that mission. The Otero Mesa and the area north of 506 (or, more accurately, the Sacramento Foothills) are only moderately utilized but play a critical role. They are also utilized for nonmilitary purposes. There is no fundamental conflict in achieving dual use as long as the Army continues existing efforts to improve interactions with nonmilitary users.

The bombing range changes the conclusions by strengthening the military arguments for renewal of the Otero Mesa. For some it raises questions about the long-term viability of dual use. At this time we do not know congressional intent regarding the need to maintain dual use. As such, we provide several policy options for Congress on the following chart.

Legislative Alternatives

Renewal Term	Modifications	Implied Results
15 years	None	Army governance
Perpetuity	None	Army governance
5 years	None	Congressional review
5 years	Public input without LEIS	Congressional review
Perpetuity	Regional Board	Cooperative governance

Figure 34

Figure 34 highlights alternative policy options that would allow Congress to implement different visions of nonmilitary use on McGregor and other 99-606 lands. The chart shows the renewal term, the modifications to 99-606, and the effects of these two factors on budgets and the issue of who determines the level of nonmilitary use.

The bombing range proposal illustrates that the current 15-year renewal is lengthy relative to the time it takes to implement projects that can degrade the level of nonmilitary use.[51] *As such, PL 99-606 establishes mechanisms for dual use, but does not ensure its existence.* As noted by the arguments in the appendix, Congress actually gave the military authority to implement projects that might erode dual use. If Congress is still content with this situation and still willing to

---

[51]The ACC Operational Impact Statement indicates that the agreement to bring the German air force to Holloman was signed in 1994, making the time to bring the Otero proposal to the point of a ROD less than four years.

allow the Army to determine the long-term uses of the range (subject to the National Environmental Policy Act and other relevant environmental statutes), then simple renewal of 99-606 for another 15-year term may be warranted. This is the first option highlighted in Figure 34. Also shown is the approximate annualized cost (total cost spread over 15 years) of complying with the 99-606 renewal process for the McGregor Range. Taken together, the six EISs for the 99-606 renewal will cost approximately $40 million, and an equal amount of money has probably been spent on studies that were prerequisites to the LEISs.[52] Whether costs for future LEISs will remain so high may depend on the level of controversy and the extent to which the military feels that only the most comprehensive documents would suffice.

This suggests that if Congress is content with Army governance and concerned about the costs of the renewals, it may want to invoke a longer renewal period. This would also eliminate the problem of the military deferring important projects for fear the land might be lost. One possibility is to withdraw the land in perpetuity. This would, however, allow the Army to develop projects that could ultimately preclude nonmilitary uses.

If Congress is not content with Army governance, and insists upon certain levels of nonmilitary uses, then a 15-year renewal term may be too lengthy. One option is to simply shorten the renewal period, as highlighted in the third option. This would prevent significant new uses from reaching approval prior to a congressional review of withdrawal status. It would, however, consume additional congressional time and DoD resources and might prevent the military from utilizing the land. Another possibility is to develop a process that mandates public input, contains a short renewal period, but does not

---

[52]The Air Force Office of Ranges and Airspace on the air staff estimates that the Goldwater and Nellis renewals cost approximately $15 million each. About half the funds were for the EIS itself and the remainder for other environmental projects needed as prerequisites for the EIS. Some Air Force officials feel the costs of the latter category may be even higher. These include GIS systems, archaeological surveys, and other items. Similarly, Fort Bliss needed to prepare a Mission and Master Plan EIS, as it was technically out of compliance with NEPA. Technically, some of these extra items were necessary even without the LEIS, some were accelerated due to the 99-606 process, and others were probably extraneous. These related items make it difficult to assess the true costs of 99-606.

require an expensive LEIS. This would cut costs and still maintain strict congressional oversight. Finally, there is a possibility of combining a long renewal term with a regional board composed of both military and nonmilitary representatives. Such a board would be empowered to halt activities inconsistent with legislative goals. Congress would need to decide how such a board would be created and how the balance of power should be determined. Although still a vague idea, it is an approach that allows joint governance without significant expense. The obvious issue, which Congress could choose to decide, is the relative balance between military and nonmilitary board members.

In summary, the cycle time for the environmental analysis needed to implement new military usages is well below the 15-year renewal period. Although Congress is unlikely to become involved in detailed local land use priorities, it will, by either direct action or preservation of the status quo, determine long-term land uses. If it is comfortable with the Army governance, then there is a strong financial argument to lengthen the renewal period. If not, it should consider alternative means of governance that are less costly.

Finally we note that DoD could facilitate congressional review by forming an interagency planning group that could identify alternatives to the current 15-year process.

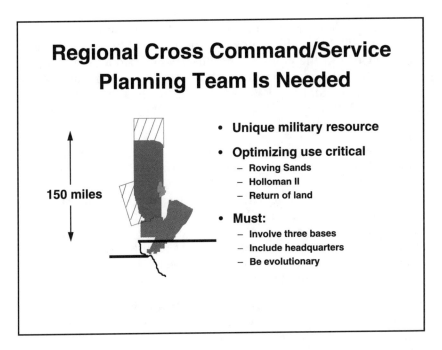

Figure 35

Figure 35 highlights an observation that emerges from several points in the report; regional military planning and optimization of land and airspace use is already a critical need and will become more urgent. A bombing range on Otero Mesa may be the best alternative for the German air force, but the lack of attention to the "existing ranges" option (and the lack of a determined effort to make that option work) weakens the argument. Roving Sands has also been damaged by the incompatibility between the financial structures of WSMR and Bliss. As noted in our analysis, the prospect of eventually returning land to the public domain may depend on military users being comfortable with the ability to perform tasks at either installation.

The need for extended parcels of military lands is likely to increase with the continual improvement of weapon system range and sensors. There will be a growing need for cross-installation land and airspace use. The ATACMS, Roving Sands, and THAAD already provide examples of how this flexibility will be beneficial. A lengthy process to approve missile firings from Fort Wingate in northern New Mexico to WSMR and Bliss provides another example of such needs. Of all regions in the country, the Bliss/WSMR/Holloman complex may contain the most unique military resource. It is almost 150 miles long. An additional 40 miles to the north, and a parcel to the west of WSMR, can be accessed on a temporary basis. These are known as call up areas.[53]

Any planning efforts would need to involve the three installations and the Department of the Army and perhaps OSD. Efforts to use WSMR for training or operational testing (as opposed to developmental testing) will immediately lead to fees that seem unrealistic to the two mission-funded installations. Headquarters needs to be present to mediate discussions, determine if there are ways to work around the financial obstacles, or at least report and document the problems as part of a long-term effort to facilitate more effective use of DoD's limited natural resource base.

Joint use is also hindered by the different missions and associated scheduling philosophies. There are different installation support requirements and different operating tempos. There must be a deliberate effort to make personnel from all three installations aware of each other's needs. It might also be desirable to schedule some demonstration activities to help determine the scope of scheduling problems. Such activities could fulfill real needs but also help identify and overcome institutional obstacles while building a tradition for joint use. One possibility is to conduct some ADA Patriot firings at WSMR to determine if there is any lost value or flexibility.

---

[53]WSMR has an agreement with the approximately 60 ranchers in the region north and west of the base that allows the area to be accessed 12 times each year. Ranchers must evacuate for 12 hours. For this, WSMR pays the ranchers a total of approximately $750,000 per year, distributed by the amount of land owned, plus a per diem. There were only four evacuations in FY 97. Recently the *Washington Post* (October 18, 1998, p. A3) included a feature article on the evacuation procedures. The article discussed the per diem but did not mention the fixed fee.

CONCLUSIONS

We conclude by presenting answers to the four key policy questions.

- Is the McGregor Range a critical Army-wide priority?

  We find that Fort Bliss has a critical role as the national center for air defense and that McGregor Range is essential for fulfilling that role.

- How intensely does the military use the McGregor Range?

  There is intense military use in the Tularosa Basin. There is a moderate level of low-impact use on the Otero Mesa and the Sacramento Foothills.

- Are military and nonmilitary uses balanced effectively, and what could change that balance?

  There are no fundamental obstacles to dual use for today's mission. The military uses have low impact and the nonmilitary uses are small. Ongoing Army efforts to respond to outside users will eliminate most conflicts. The new bombing range on Otero Mesa will not affect this, but it has created political concerns about the future. However, Congress clearly intended to give the military services the discretion to initiate projects that would affect dual use. Nevertheless, the dual use priority has discouraged Army interest in utilizing the Otero Mesa for new military applications.

- Is it possible to transfer McGregor activities to the adjacent White Sands Missile Range, which has more land and declining activity?

  There is probably sufficient land and airspace to transfer most activities on Otero Mesa and the area north of 506 to WSMR. The situation may not be ideal for the Roving Sands exercise and would force other Fort Bliss units to travel greater distances. Substantial DoD policies and procedures currently prevent consideration of this transfer.

We also recommend that DoD establish a regional planning committee to facilitate joint use of the Holloman/WSMR/Bliss complex. This must be composed of representatives from the three installations and from higher headquarters.

LEGISLATIVE HISTORY OF PUBLIC LAW 99-606

Donald Mitchell

OVERVIEW

The potential development of a bombing range on the Otero Mesa raises the question of congressional intent regarding the nonmilitary uses of lands withdrawn under Public Law 99-606.[1] While the new bombing range comprises only 2 percent of the mesa, it is seen by some as the type of project that could ultimately lead to erosion of the nonmilitary uses.

This appendix reviews the legislative history of 99-606 with the purpose of determining congressional intent. The statutory text and legislative history of the Act indicate that the 99th Congress intended the Military Lands Withdrawal Act (MLWA) to delegate to the secretary of the military department concerned the authority to unilaterally allow the exclusive use for defense-related purposes of public land, and all portions thereof, that have been withdrawn by section 1 of MLWA, other than public land in the Nellis and Barry M. Goldwater Air Force Ranges that is located within the Desert and Cabeza Prieta National Wildlife Refuges. Public land located within the two refuges may not be used for defense-related purposes:

1. unless the defense-related purpose was authorized in a memorandum of agreement in existence on the date of enactment of MLWA, or

---

Donald Mitchell is an attorney in Anchorage, Alaska. He is a member of the bar of the Supreme Court of the United States and former General Counsel of the Alaska Federation of Natives.

[1]Pub. L. No. 99-606. 100 Stat. 3457 (1986).

2. with respect to a defense-related purpose that was not authorized in a memorandum of agreement in existence on the date of enactment of MLWA, the Secretary of the Interior determines that the use of refuge land for such purpose will be compatible with the major purposes for which the refuge was established.

Management of Non-National Wildlife Refuge MLWA Withdrawn Land

In response to the exigencies of war, beginning in 1940 the President (by executive order) and the Secretary of the Interior (by public land order) withdrew several million acres of public land in several western states and Alaska for defense-related purposes. When the Second World War ended and the Cold War began, the Department of Defense (DoD) requested the Secretary of the Interior to make additional withdrawals. Numerous withdrawals were made. But by 1955 requests for withdrawals that collectively totaled more than four million acres of public land still were pending.[2] Ranchers who leased grazing rights, mining companies, recreational hunters and fishermen, and other individuals and interests who desired to use the public land that DoD had asked the Secretary of the Interior to withdraw vociferously objected to "the Department of Defense . . . by a single stroke of the pen tak[ing] so much of the public domain] . . . out of the multiple-use category" for a "single purpose [i.e., military] use."[3] And in response to those objections, in 1958 the 85th Congress enacted the Engle Act.[4]

The Engle Act provided *inter alia* that "except in time of war or national emergency" no withdrawal of public land "for the use of the Department of Defense for defense purposes" that in the aggregate totaled five thousand or more acres could be made except by Act of Congress. Significantly, the 85th Congress not only did not intend

---

[2]See Hon. Clair Engle, chairman, Committee on Interior and Insular Affairs, U.S. House of Representatives, to Charles E. Wilson, Secretary of Defense, December 27, 1955, reprinted in *Withdrawal and Utilization of the Public Lands of the United States: Hearings before the House Committee on Interior and Insular Affairs,* 84th Cong., 2d Sess. (1956) [hereinafter "1956 Hearings"].

[3]1956 Hearings, p. 3.

[4]Pub. L. No. 85-337, 72 Stat. 27 (codified at 43 U.S.C. 155-58)(1958).

the Engle Act to affect the status and management of public land that the President and the Secretary of the Interior previously had withdrawn for defense purposes, but section 1(4) of the Act explicitly provided that the provisions of the Engle Act did not affect "reservations or withdrawals which expired due to the ending of the unlimited national emergency of May 27, 1941, and which subsequent to such expiration have been and are now used by the military departments with the concurrence of the Department of the Interior . . . ."

By 1979 there were seven areas in five states—California, Nevada, New Mexico, Arizona, and Alaska—which in the aggregate collectively encompassed approximately 7.51 million acres of public land that had been withdrawn for defense-related purposes by executive or public land order or statute, whose withdrawals had expired, and whose occupation by DoD military departments consequently was technically unlawful:[5]

1. Mojave-B Range

2. Bravo 20 Bombing Range

3. Nellis Air Force Range

4. Luke Air Force Range

5. McGregor Range

---

[5]In 1979 the Department of the Interior Bureau of Land Management advised the House Committee on Interior and Insular Affairs that

> In the case of withdrawals that have expired, we know of no specific authority for continued occupancy during the interim, application processing period and, if appropriate, the period required for Congressional consideration of proposed withdrawal legislation. However, we believe there would be no useful purpose served in attempting to eject another Federal agency from the lands when steps are being taken to resolve the question of continued occupancy and reservation of the land.

Arnold E. Petty, acting associate director, Bureau of Land Management, to William L. Shafrer, House Interior and Insular Affairs Committee staff, April 4, 1979, reprinted at *Additions to the National Wilderness Preservation System: Hearings on H.R. 5426, 5470, 4932, 5965 and S. 837 before the Subcommittee on Public Lands and National Parks of the House Committee on Interior and Insular Affairs,* 98th Cong., 2d Sess. 230 (1984) [hereinafter "1984 House Hearings"].

6. Fort Greely Maneuver Area and Air Drop Zone

7. Fort Wainwright Maneuver Area

To rectify the problem, officials and attorneys representing DoD and the Department of the Interior Bureau of Land Management (BLM) jointly drafted seven bills whose enactment would have withdrawn the seven areas for twenty-five years. In 1984 the Department of the Interior sent the bills to the 98th Congress where they were introduced in the Senate by Senator James McClure, the chairman of the Committee on Energy and Natural Resources, and in the U.S. House of Representatives by Representative Melvin Price, the chairman of the Committee on Armed Services.[6]

The texts of the seven bills differed in particulars such as the descriptions of the defense-related purposes for which each area would be withdrawn or the reservation of water rights. But with regard to structure and policy objectives, the texts were virtually identical. As Frank A. Edwards, the assistant director of the BLM, explained during a hearing that the House Committee on Interior and Insular Affairs Subcommittee on Public Lands and National Parks held on H.R. 4932,[7] the bill to withdraw the Nellis Air Force Range,

> The legislation under consideration today was drafted by representatives of the Bureau of Land Management in close coordination with representatives of the Department of the Air Force and the other military departments. H.R. 4932 is only one of several military withdrawal bills that have been forwarded to the Congress for action. Because all of the proposed withdrawals are similar in nature, a decision was made early in the negotiation process [between DoD and BLM] that all of the military withdrawal bills forwarded to Congress would be similar in format and language.[8]

---

[6]See S. 2656 and H.R. 6322 (Mojave-B Range), S. 2657 and H.R. 4932 (Nellis Air Force Range), S. 2658 and H.R. 4933 (McGregor Range), S. 2659 and H.R. 6319 (Luke Air Force Range), S. 2660 and H.R. 6321 (Fort Greely Maneuver Area and Air Drop Zone), S. 2661 and H.R. 6320 (Fort Wainwright Maneuver Area), and S. 2662 and H.R. 6323 (Bravo-20 Bombing Range), 98th Cong., 2d Sess. (1984).

[7]In the U.S. House of Representatives, H.R. 4932-33 and 6319-23 were jointly referred to the Committee on Interior and Insular Affairs and the Committee on Armed Services.

[8]1984 House Hearings, p. 210.

With respect to the respective authority of the secretaries of the military departments concerned and the Secretary of the Interior to manage the same withdrawn public land, the texts of the bills were ambiguous. For example, while section 2(a) of H.R. 4932 granted the Secretary of the Air Force "exclusive jurisdiction over the management of the lands [located within the Nellis Air Force Range] for military purposes," section 2(b) of the bill granted the Secretary of the Interior authority to manage the same lands "under principles of multiple use and sustained yield, for uses which may include, but are not limited to, grazing, management of wildlife habitat, control of predatory animals, and the prevention and suppression of brush and range fires resulting from nonmilitary activities."

To compound the ambiguity, section 2(a)(i) of H.R. 4932 granted the Secretary of the Air Force authority to close "roads and trails commonly in public use" within the Nellis Air Force Range whenever he determined that "military operations, public safety or national security" necessitated a closure. However, section 2(a)(i) did not grant the secretary authority to close the public land that the roads and trails crossed.

The drafters of H.R. 4932 were aware of the potential for land management conflict. To reduce it, they included provisions in the bill that mandated a planning process to develop a "resource management plan" that would be implemented by a "memorandum of understanding to implement the plan" into which the bill directed the Secretary of the Air Force and the Secretary of the Interior to enter. However, the text of H.R. 4932 provided no methodology for resolving land use disputes between the Air Force and the BLM that the two secretaries could not resolve.

In that regard, the following colloquy—which took place during the aforementioned hearing on H.R. 4932 among Representative John Seiberling (chairman of the House Committee on Interior and Insular Affairs Subcommittee on Public Lands and National Parks), John O. Rittenhouse (deputy for installations management, Office of the Deputy Assistant Secretary of the Air Force), and Frank A. Edwards (assistant director, BLM)—is instructive:

SEIBERLING: Suppose there is a disagreement, who would have the final say?

RITTENHOUSE: I believe the Secretary of the Air Force would hold some degree of leverage there, sir.

SEIBERLING: So if there is a difference, the Secretary of the Air Force would make that decision?

RITTENHOUSE: As I recall the legislation, that is the way the legislation indicates that.

SEIBERLING: It isn't clear, that is why I wanted to know. Well, I suppose in the end the President would be the final arbiter.

RITTENHOUSE: Yes, sir.

SEIBERLING: Section 2(a) of the bill provides the Secretary of the Air Force shall have exclusive jurisdiction over the management of the lands for military purpose [sic] and may authorize use by other military departments and agencies of the Department of Defense, Department of Energy, as appropriate. Now that is a little different from who decides what is rendered necessary by national defense, isn't it? Because the rendered necessary by national defense involves making a decision between military and nonmilitary activities, which I would think is not covered by section 2(a), is that right?

EDWARDS: Well, it is my understanding that the Secretary of the Air Force would have the exclusive jurisdiction for all military operations; those by the Air Force as well as those by other military branches. And again it would be a consultative process on the other, but in fact, the Air Force would have the final say subject to the approval of the two Secretaries and the President.

SEIBERLING: My question is, do the words "for military purposes" imply that if there is a question about whether something is rendered necessary for purposes of section 1(a), the Secretary of Defense has that final say-so or the Secretary of the Air Force?

EDWARDS: Yes; it could be read that way. But again, as we have indicated, the full intent is to have the consultative process.

SEIBERLING: I think we had better make that clear somehow.

RITTENHOUSE: If you look at 2(b), I think maybe the sequence may be Secretary of Interior shall manage withdrawn lands and their resources. That is where the memorandum of agreement comes into play. Because in (c) it indicates that there shall be a memorandum of understanding to implement the plan, and that is exactly what we do have now, in effect.

EDWARDS: The bill does require that a management plan be developed within one year and, of course, that plan could lay out clearly how disputes or disagreements will finally be resolved.[9]

Subsequent to the hearing, the members of the Subcommittee on Public Lands and National Parks decided not to take action on H.R. 4932 and the six companion bills during the 98th Congress. Instead, they decided to deal with the military lands withdrawal issue during the 99th Congress. But before leaving the subject, the Subcommittee wrote, the Committee on Interior and Insular Affairs reported, and the 98th Congress enacted, an amendment in the nature of a substitute for the original text of H.R. 4932 that temporarily withdrew the Groom Mountain Range, an 89,600-acre tract of public land located proximate to the Nellis Air Force Range within which the Air Force needed authority to prohibit public access for reasons of national security.[10]

In two important respects the text of the amendment in the nature of a substitute resolved the ambiguity that Representative Seiberling had identified during the aforementioned hearing on H.R. 4932 regarding the authority of the Secretary of the Air Force to prohibit the Secretary of the Interior's multiuse management in situations in

---

[9]1984 House Hearings, pp. 227–228.

[10]See Pub. L. No. 98-485, 98 Stat. 2261 (1984).

which the Secretary of the Air Force determined that a tract of public land was needed exclusively for a defense-related purpose.

First, section 3(a) of the amendment provided that while the Secretary of the Interior was authorized to manage public land within the Groom Mountain Range withdrawal pursuant to the Federal Land Policy and Management Act (FLPMA), all FLPMA-related non-defense-related uses "shall be secondary to the military use of such lands" and "may be authorized by the Secretary of the Interior only with the concurrence of the Secretary of the Air Force."

Second, section 3(b) of the amendment rewrote the text of the section of H.R. 4932 that granted the Secretary of the Air Force authority to close roads and trails to also grant the secretary authority to "clos[e] to public use . . . any other portion of the lands withdrawn by this Act."

Those provisions clarified that during the period of the temporary withdrawal of the Groom Mountain Range, the Secretary of the Air Force had authority to unilaterally decide (without having to obtain the concurrence of the Secretary of the Interior) that the range, or any portion thereof, should be used exclusively for defense-related purposes and to close the range, or any portion thereof, to all non-defense-related uses.

In 1985, when the 99th Congress convened, the House Committees on Interior and Insular Affairs and Armed Services decided to consolidate the withdrawals that the Secretary of the Interior had sent to the 98th Congress as separate bills into a single bill. And in March 1985, Representative Beverly Byron, who was a member of both committees, introduced the new bill in the U.S. House of Representatives as H.R. 1790.[11]

---

[11]The original text of H.R. 1790 is reprinted at 132 Cong. Rec. 33101-03 (1986). H.R. 1790 did not include the withdrawal of the Mojave "B" Range on California that had been introduced as H.R. 6322/S. 2656 in the 98th Congress. According to Stanley Sloss, who during the 99th Congress was counsel to the Subcommittee on Public Lands and National Parks, the Mojave "B" Range was omitted from H.R. 1790 because before Representative Byron's introduction of the bill, representatives of the Department of the Navy failed to inform her of the need to withdraw the range. Telephone communication from Stanley Sloss to Donald Mitchell, August 7, 1998.

Neither committee reported H.R. 1790 during the first session of the 99th Congress. Displeased with the delay, at the beginning of the second session Representative Barbara Vucanovich of Nevada in March 1986 introduced H.R. 4351, a bill whose enactment would have withdrawn the Nellis and Bravo 20 ranges in Nevada. And in August, Representative Don Young of Alaska introduced H.R. 5389, a bill whose enactment would have withdrawn the Fort Wainwright Maneuver Area and Fort Greely Maneuver Area and Air Drop Zone. In July and September 1986 respectively, the Committee on Interior and Insular Affairs reported amendments in the nature of substitutes for the original texts of H.R. 4351 and H.R. 5389 to the U.S. House of Representatives.[12]

Both amendments in the nature of substitutes attempted to reduce the potential for land management conflict by requiring the Secretary of the Interior (after consultation with the secretary of the military department concerned) to prepare a management plan for each area and then requiring both secretaries to enter into a memorandum of understanding regarding implementation of the plan. And section 4(b)(1) of both amendments adopted the closure standard that had been included in section 3(b) of the amendment in the nature of a substitute for H.R. 4932 (the bill that temporarily withdrew the Groom Mountain Range), which had been enacted by the 98th Congress. As mentioned, in addition to roads and trails, that provision authorized the secretary of the military department concerned to unilaterally close "any other portion" of the withdrawn land to nonmilitary uses whenever he determined that "military operations, public safety, or national security" required that he do so.

Although they shared those commonalities, the two amendments differed quite dramatically regarding the extent to which the secretary of the military department concerned would be authorized to use a tract of withdrawn public land for a defense-related purpose other than the specific defense-related purposes listed in section 2 of both amendments.

Section 2(a)(1)(C) and (b)(1)(B) of the amendment to H.R. 4351 authorized (but did not require) the Secretary of the Interior (rather

---

[12]See H. Rep. Nos. 99-689 and 920 (1986).

than the secretary of the military department concerned) to permit public land within the Nellis and Bravo 20 ranges to be used for other defense-related purposes. And section 4(g)(2) of the amendment directed the Secretary of the Interior to authorize public land within the ranges to be used for such other defense-related purposes *only* in situations in which he determined that the use of the land for such other defense-related purposes would not (either alone or in combination with the specified defense-related purposes) "have a significant impact on the resources and values of the affected lands," and would not result in "the imposition of additional or more stringent conditions or restrictions on otherwise permitted nonmilitary uses of the affected lands than are required to permit the uses specified in section 2."

By contrast, section 2(a)(1)(C) and (b)(1)(B) of the amendment to H.R. 5389 reserved to the Secretary of the Army (rather than the Secretary of the Interior) the decision on whether public land within the Fort Wainwright Maneuver Area and Fort Greely Maneuver Area and Air Drop Zone should be used for defense-related purposes other than those specified in section 2 of the amendment. And section 4(g) of the amendment simply required the Secretary of the Army to "notify the Secretary of the Interior in the event that the lands withdrawn by this Act will be used for defense-related purposes other than those specified in section 2."

In August 1986 the U.S. House of Representatives passed the amendment in the nature of a substitute for the original text of H.R. 4351 without controversy and after only cursory debate.[13] And in September 1986 the House passed the amendment in the nature of a substitute for the original text of H.R. 5389, again without controversy and after only cursory debate.[14]

In the meantime, in the Senate, by the second session of the 99th Congress, Senator James McClure, the chairman of the Committee on Energy and Natural Resources, also had decided that the military withdrawals should be made in a single bill. In May 1986 Senator

---

[13] 132 Cong. Rec. 21052-59 (1986).
[14] 132 Cong. Rec. 27355-57 (1986).

McClure introduced S. 2412 as a companion bill to H.R. 1790.[15] In July the Committee on Energy and Natural Resources Subcommittee on Public Lands, Reserved Water and Resource Conservation, to which S. 2412 had been referred, held a hearing on the bill,[16] and in October the Committee reported the original text of S. 2412 (and five amendments thereto) to the Senate.[17]

S. 2412 resolved the potential for conflict between the secretaries of the military departments concerned and the Secretary of the Interior regarding defense-related and non-defense-related uses of the withdrawn public land in favor of the secretaries of the military departments concerned. The bill did so by in section 1 denying the Secretary of the Interior any authority to participate in the authorization of public land within the withdrawn areas for use for a defense-related purpose not specifically identified in section 1 for each area. And section 4 of the bill granted the secretaries of the military departments concerned exclusive authority "to control the military use of the lands."[18]

After establishing the primacy of the authority of the secretaries of the military departments concerned, S. 2412 attempted to prevent land use conflicts by in section 3(b)(1) requiring the Secretary of the Interior and the secretaries of the military departments concerned to jointly develop "a land use plan and management program for the use and management of the lands withdrawn and reserved by this Act." And section 4(e) directed the secretaries to "enter into a memorandum of understanding to implement the program."

In broad concept, the amended version of S. 2412 that the Senate Committee on Energy and Natural Resources reported differed from

---

[15]In addition to the five withdrawals contained in H.R. 1790, S. 2412 also withdrew the Chocolate Mountain Aerial Gunnery Range and China Lake Naval Weapons Center. The text of S. 2412 is reprinted at 132 Cong. Rec. 9665-67 (1986).

[16]*Land Withdrawals from the Public Domain for Military Purposes: Hearing on S. 2412 before the Subcommittee on Public Lands, Reserved Water and Resource Conservation of the Senate Committee on Energy and Natural Resources*, 99th Cong., 2d Sess. (1986).

[17]S. Rep. No. 99-514 (1986).

[18]While section 4(b) granted the secretaries of the military departments concerned authority to close "roads and trails commonly in public use" when "military operations, public safety or national security" required, section 4(b) did not grant the secretaries authority to close public land that the roads and trails crossed.

the version of H.R. 1790 that was pending in, but which had not been reported by, the House Committee on Interior and Insular Affairs in that:

1. H.R. 1790 did not withdraw the Chocolate Mountain Aerial Gunnery Range and China Lake Naval Weapons Centers, both of which areas were included in S. 2412.

2. The withdrawals in H.R. 1790 expired ten years after the date of enactment of MLWA, while (as DoD had recommended in 1984) the withdrawals in S. 2412 expired twenty-five years after the date of enactment of MLWA.

3. S. 2412 renamed the Luke Air Force Range as the Barry M. Goldwater Air Force Range.

To resolve those differences, a member of the staff of the House Committee on Interior and Insular Affairs and a member of the staff of the Senate Committee on Energy and Natural Resources conducted an informal negotiation.[19] The outcome of the negotiation was as follows:

1. The Senate Committee on Energy and Natural Resources negotiator agreed to abandon S. 2412 and accept the text of the amendment in the nature of a substitute for H.R. 1790 that had been written, but not yet reported, by the House Committee on Interior and Insular Affairs.[20]

2. The House Committee on Interior and Insular Affairs negotiator agreed to add a provision to the amendment in the nature of a substitute for H.R. 1790 that renamed the Luke Air Force Range as the Barry M. Goldwater Air Force Range.

3. Both negotiators agreed that, rather than ten or twenty-five years, the withdrawals made in the amendment in the nature of a substi-

---

[19]Donald Mitchell interview with Stanley Sloss, July 14, 1998.

[20]Since H.R. 1790 did not withdraw the Chocolate Mountain Aerial Gunnery Range and China Lake Naval Weapons Center, those withdrawals were not included in MLWA. As a consequence, the Chocolate Mountain and China Lake areas were not withdrawn until 1994, when the 103d Congress enacted the Military Lands Withdrawal and Overflights Act as part of the California Desert Protection Act. See Pub. L. No. 103-433, Title VIII, 108 Stat. 4471, 4501 (1994).

tute for H.R. 1790 would expire fifteen years after the date of enactment of MLWA.

The members of the two committees approved the negotiated agreement. The Senate and U.S. House of Representatives then without controversy enacted MLWA into law as an amendment in the nature of a substitute for the original text of H.R. 1790.[21]

When read collectively, the text of four sections of the amendment in the nature of a substitute for the original text of H.R. 1790 indicates that the 99th Congress intended MLWA to grant the secretaries of the military departments concerned authority to (without obtaining the concurrence of the Secretary of the Interior) (1) close any or all portions of public land within a withdrawn area to any or all non-defense-related uses; and (2) use any or all portions of public land within a withdrawn area (other than land located within a national wildlife refuge) for a defense-related purpose that was not specified in the subsection of section 1 of MLWA that withdrew the area.

The four sections are:

Section 3(b)(1)

Provides that "If the Secretary of the military department concerned determines that military operations, public safety, or national security require the closure to public use of any road, trail, *or other portion* of the lands withdrawn by this Act, the Secretary may take such action as the Secretary determines necessary or desirable to effect and maintain such closure" (emphasis added).[22]

Section 3(a)(3)(B)

Prohibits the Secretary of the Interior from leasing, or granting an easement, right-of-way, or other authorization with respect to

---

[21]The U.S. House of Representatives passed the negotiated version of H.R. 1790 on October 17, 1986. 132 Cong. Rec. 33100-07 (1986). The Senate did the same on October 18, 1986. 132 Cong. Rec. 33819-20 (1986).

[22]Section 3(a)(2) of MLWA authorizes the Secretary of the Interior to continue grazing and recreation within MLWA withdrawal areas. However, section 3(b)(1) authorizes the secretaries of the military departments concerned to close any "portion of the lands withdrawn" to grazing, recreation and other nonmilitary uses.

the nonmilitary use of public land located within a MLWA withdrawal area unless he obtains "the concurrence of the Secretary
of the military department concerned."

Subparagraphs in each Subsection in Section 1

Provide that, in addition to the specific defense-related purposes
described in each subsection of section 1 that withdraws each
MLWA withdrawal area, each withdrawn area is reserved for use
by the secretary of the military department concerned for other
defense-related purposes consistent with the specific purposes
as long as the secretary complies with the provisions of section
3(f).

Section 3(f)

Provides that "Lands withdrawn by section 1 (except those
within the Desert National Wildlife Range [sic] or within the
Cabeza Prieta National Wildlife Refuge) may be used for defense-
related uses other than those specified in such section. The Secretary of Defense shall promptly notify the Secretary of the Interior in the event that the lands withdrawn by this Act will be used
for defense-related purposes other than those specified in section 1. Such notification shall indicate the additional use or uses
involved, the proposed duration of such uses, and the extent to
which such additional military uses of the withdrawn lands will
require that additional or more stringent conditions or restrictions be imposed on otherwise-permitted nonmilitary uses of the
withdrawn land or portions thereof."[23]

---

[23]Compare the notification requirement in section 3(f) with the provisions of the
version of H.R. 4351, the bill to withdraw the Nellis Air Force Range and Bravo-20
Bombing Range, that the U.S. House of Representatives passed two months before it
passed MLWA. As previously described, section 2(a)(1)(C) and (b)(1)(B) of H.R. 4351
authorized the Secretary of the Interior (rather than the secretary of the military
department concerned) to permit public land within the ranges to be used for other
defense-related purposes. And section 4(g)(2) directed the Secretary of the Interior to
authorize public land within the ranges to be used for other defense-related uses *only*
in situations in which he determined that the other uses would not (either alone or in
combination with the specified defense-related uses) "have a significant impact on the
resources and values of the affected lands," and would not result in "the imposition of
additional or more stringent conditions or restrictions on otherwise permitted nonmilitary uses of the affected lands than are required to permit the uses specified in
section 2."

Management of National Wildlife Refuge Public Land Withdrawn Under 99-606

The boundary of the Nellis Air Force Range overlaps the western half of the Desert National Wildlife Refuge. And the boundary of the Barry M. Goldwater Air Force Range overlaps a portion of the Cabeza Prieta National Wildlife Refuge. Within the overlap areas, the 99th Congress intended MLWA to establish a land management system different from the system described above.

Section 3(b)(1) of MLWA authorizes the Secretary of the Air Force to close any or all portions of the overlap areas to any or all non-defense-related uses whenever he determines that "military operations, public safety, or national security" so require. However, with respect to the secretary's use of public land within the overlap areas for defense-related purposes, MLWA divides defense-related purposes into two categories.

In 1966, the 89th Congress enacted the National Wildlife Refuge Administration Act (NWRAA).[24] NWRAA consolidated all public land administered by the Secretary of the Interior for the conservation of fish and wildlife into a National Wildlife Refuge System, and established uniform standards for the management of land and natural resources within the system. The most important management standard, which is contained in section 4(d)(1) of NWRAA, authorizes the Secretary of the Interior to "permit the use of any area within the [National Wildlife Refuge] System for any purpose . . . whenever he determines that such uses are *compatible* with the *major purpose* for which such areas were established" (emphasis added).

In 1936 President Franklin Roosevelt by executive order established the Desert National Wildlife Range for the purpose of the "conservation and development of natural wildlife resources and for the protection and improvement of public grazing lands and natural forage resources."[25] And in 1939 President Roosevelt established the Cabeza Prieta National Wildlife Range for the same purpose.[26]

---

[24]Pub. L. No. 89-669, 80 Stat. 927 (codified at 16 U.S.C. 668dd-jj)(1966).

[25]Executive Order No. 7373 (May 20, 1936).

[26]Executive Order No. 8038 (Jan. 25, 1939).

When officials and attorneys representing DoD negotiated with their counterparts at the BLM regarding the text of the seven military land withdrawal bills that the Department of the Interior in 1984 sent to the 98th Congress, they were concerned that the defense-related uses that were occurring within the areas of the Nellis and Luke Air Force Ranges that overlapped the Desert and Cabeza Prieta National Wildlife Refuges were not "compatible" with the purpose for which the refuges had been established. If they were not, the Secretary of the Interior's authorization of such uses was unlawful.[27]

To remedy the problem, section 2(b) of S. 2657/H.R. 4932 and S. 2659/H.R. 6319, the bills to withdraw the Nellis and Luke Air Force Ranges, stated that public land located within the Desert and Cabeza Prieta National Wildlife Refuges was to be managed by the Secretary of the Interior in accordance with NWRAA "except to the extent rendered necessary by national defense."

In other words, public land located within the refuges could be used for a defense-related purpose even if the use was incompatible with the purpose for which the refuges had been established if it was determined that the nation's defense made doing so necessary. The statute is not clear as to which department has the authority to make that determination.

Sections 3(b)(2) and (3) of S. 2412, the consolidated bill that Senator James McClure introduced in 1986, contained a national defense exception to the NWRAA compatibility test. However, neither the original text of H.R. 1790 nor the amendment in the nature of a sub-

---

[27]DoD's concern had a substantial basis in fact. In 1966 when it reported H.R. 9424, the bill that the 89th Congress enacted as NWRAA, the House Committee on Merchant Marine and Fisheries directed the Secretary of the Interior "to be very cautious in permitting compatible uses" and to authorize uses of public land within the National Wildlife Refuge System "only when extreme caution has been exercised to make sure that the other uses are compatible and incidental and secondary to the primary purposes." H.R. Rep. No. 89-1168, pp. 8, 11 (1966). If the Secretary of the Interior's authority to manage public land located within the Desert and Cabeza Prieta National Wildlife Refuges had been subject to the NWRAA compatibility standard at the time he and the Secretary of the Air Force negotiated the first memoranda of understanding that authorized land within the refuges to be used for defense-related purposes, it is quite likely that the Secretary of the Interior would have determined that dropping ordnance on public land that had been withdrawn for the purpose of protecting and improving grazing land and natural forage would be inconsistent with the achievement of that purpose.

stitute for the original text of H.R. 1790 that the 99th Congress enacted as MLWA contained a national defense exception. As a consequence, sections 4(a)(1)(A) and (b)(1)(A) of MLWA simply state that neither section 1 (which withdraws the Nellis and Barry M. Goldwater Air Force Ranges) nor any other provision of MLWA "shall be construed to amend the National Wildlife Refuge Administration Act of 1966 or any other law related to management of the National Wildlife Refuge System."[28]

Although MLWA does not contain a national defense exception to the compatibility test, sections 4(a)(2) and (b)(2) of MLWA state that the provisions of MLWA do not amend the memoranda of understanding regarding the joint use of the Desert and Cabeza Prieta National Wildlife Refuges into which the Secretary of the Air Force and the Secretary of the Interior had entered and which were, in effect, prior to the enactment of MLWA.

As a consequence, the use of public land located within the Desert and Cabeza Prieta National Wildlife Refuges for defense-related pur-

---

[28]The 99th Congress's refusal to create a national defense exception to the NWRAA compatibility test was not unprecedented. The original text of H.R. 9424, the bill the 89th Congress enacted as NWRAA, was drafted by attorneys at the Department of the Interior. See Stewart L. Udall, Secretary of the Interior, to the Hon. John W. McCormick, Speaker, U.S. House of Representatives, June 5, 1965, reprinted at H.R. Rep. No. 89-1168, pp. 12-14 (1966). Section 4(d)(2) of the original text authorized the Secretary of the Interior to grant easements for power and telephone lines, canals, pipelines and roads across public land located within a unit of the National Wildlife Refuge System even if doing so would be incompatible with the major purposes for which the unit had been established if the secretary determined that granting the easement would be in the "public interest." When it reported H.R. 9424 to the U.S. House of Representatives, the Committee on Merchant Marine and Fisheries recommended that the "public interest" exception to the compatibility test be eliminated. The Committee (H.R. Rep. No. 89-1168, p. 11 (1966)) advised the House that

> In view of the pressures that may be brought to bear on the Secretary, your committee was fearful that to allow the Secretary to have this discretionary authority to permit uses of the areas within the system that are incompatible, far too many incompatible uses would be found compatible and otherwise in the national interest. Therefore, your Committee appropriately deleted this provision.

The 89th Congress accepted the Committee on Merchant Marine and Fisheries' recommendation. If the 99th Congress had included a national defense exception to the NWRAA compatibility test in MLWA, doing so would have created an exception that, like the public-interest exception that the 89th Congress rejected, would have obviated the test.

poses that were authorized in the memoranda of understanding may continue, regardless of whether the use of the land for those purposes is compatible with the purpose for which the refuges were established.[29] However, the Secretary of the Interior must authorize the use of the land for a defense-related purpose not listed in the memoranda of understanding. And pursuant to section 4(d)(1) of NWRAA, the Secretary of the Interior may not authorize use for a new defense-related purpose unless he (rather than the Secretary of the Air Force) determines that the use will be compatible with the purpose for which the refuge was established. And if the Secretary of the Interior determines that a proposed new use would be incompatible, he may not authorize the use, *even if the incompatible use is necessary for national defense.*

---

[29]For example, the memorandum of understanding regarding the use by the Department of the Air Force of public land within the Desert National Wildlife Refuge authorizes the Air Force to establish training and testing facilities within the range, deliver air-to-ground ordnance within designated impact areas, and install receiver/transmitter systems. A copy of the memorandum is reprinted at H.R. Rep. No. 98-1046, Part I, pp. 10–15.